NO STONE UNTURNED

NO STONE UNTURNED

Reasoning About Rocks and Fossils

E. K. Peters

W. H. Freeman and Company
New York

ACQUISITIONS EDITOR: Holly Hodder
PROJECT EDITOR: Christine Hastings
TEXT AND COVER DESIGNER: Vicki Tomaselli
COVER PHOTOGRAPHER: John S. Shelton
TEXT DESIGNER: Edward Smith Design, Inc.
ILLUSTRATION COORDINATOR: Bill Page
ILLUSTRATOR(S): Hadel Studio
PRODUCTION COORDINATOR: Sheila E. Anderson
COMPOSITION: Susan Cory/W. H. Freeman Electronic Publishing Center
MANUFACTURING: Maple Vail

Library of Congress Cataloging-in-Publication Data

Peters, E. K.
No stone unturned: reasoning about rocks and fossils / E. K. Peters
p. cm.
Includes bibliographical references and index.
ISBN 0-7167-2953-9 (soft cover)
1. Geology—Philosophy. I. Title.
QE6.P48 1996
550.1—dc20 96-21794
CIP

Printed in the United States of America

Third printing, 2003

The second printing of this book
is dedicated to the memory of

Sheldon Judson

who was both a fine teacher
amd a true friend.

CONTENTS

PREFACE

Science is one of the liberal arts, a fact often forgotten in the rush toward professionalism. All too frequently, undergraduates in science classes are taught that graduate work is the main goal of a university education. Although many of our colleagues in the humanities and social sciences continue to view their mission in broad terms, we scientists often have retreated to narrow specialization, even in lower-division teaching.

Fortunately, however, the National Science Foundation has recently recommended that certain aspects of the humanities also be made part of introductory science instruction. More important, some colleges and universities are beginning to require that their science departments offer interdisciplinary classes—what one might call the scientific equivalent of World Civilization courses. These developments should help make it clear that although science may be one of our society's greatest and most prestigious intellectual accomplishments, it is not divorced from other types of thinking, including nonempirical disciplines. Accordingly, undergraduates need to "get back to the basics" of the college curriculum, at least for a semester or two, to understand how different branches of the liberal arts—including the sciences—complement one another.

The geological sciences are uniquely qualified to be the leader in this undertaking. Some parts of geology use the content and methodology of the physical sciences; other parts have devised their own procedures to cope with the interpretive and historical aspects of the Earth sciences. In short, because geology uses the whole continuum of scientific methodology, it is able to expose students to different types of empirical inquiry.

No Stone Unturned was not written in an idealistic vacuum, but is based on my experiences in the interdisciplinary science class that I teach. Introducing the newcomer to the range and quality of scientific debate is this book's goal, using such questions as

- What is the scope and nature of modern science?
- How does historical science differ from physical science?
- How is science related to nonempirical knowledge?

Rather than being a dreary journey through geological vocabulary, this book examines the structure and implications of major scientific disputes in their historical contexts. The examples used here include early ideas about continental drift, contemporary problems in tectonic theory, the possible origins of life in the Hadean eon, the catastrophic flooding events of the Pleistocene in the Pacific Northwest, and current arguments regarding evolutionary theory, especially with respect to the known fossil record. These topics, I have found, are interesting to nonscience majors and science majors alike.

Although this book can stand on its own, it can also be used as a supplement to a traditional introductory textbook, particularly for discussions of the scientific method and the concept of uniformitarianism, which are mentioned but seldom developed in most Geology 101 classes. And although this book is certainly not too difficult for freshmen, upperclass geology majors and graduate students may find it helpful as well. Technical terms are defined when they first appear and are also in the glossary. Each chapter ends with questions, many of which could be used on essay exams or as starting points for in-class debates. The questions are designed to make students think critically and to practice discussing the abstract ideas on which geology rests. Recommendations for further reading can be found in the annotated bibliography at the end of the book.

A final point: I hope that this discussion of modern science is of interest not only to students but also to those responsible members of our society who spend their days in the working world. In several respects, these "interested laypersons" were uppermost in my mind when I wrote this book.

ACKNOWLEDGMENTS

This project has been a pleasure from beginning to end. A large part of my enjoyment has come from responding to the ideas and criticisms of two professors who, long ago, taught me introductory physical geology and the history of life on earth. Sheldon Judson (Princeton) patiently read and reread these pages and found many of my errors while simultaneously reorganizing this material along more rational lines. Alfred Fischer (formerly of Princeton, now at the University of Southern California) successfully urged reconsideration on several points where I had wandered into unnecessary cynicism. Few students have a chance to lecture their own teachers about methodological limitations and pedagogical error, and few teachers would respond so well as mine.

My colleagues at Washington State University have assisted me at several different levels. I am in debt to the following professors: Jerry Gough, John Thompson, Robert Jonas, Kent Keller, and Walfred Peterson. Matthew Turnbull, a former W.S.U. student, read the first draft and respectfully recommended clarity where I had produced obfuscation. Finally, Larry Davis, now at St. John's University, was not only brave enough to team-teach with me, he also encouraged me at each step of the writing and photographic processes. My job, and therefore the existence of this book, flow directly from Larry's ability to direct me toward useful work.

Reviewers for W. H. Freeman and Company have pointed out errors and added strength to some of the weaker arguments. I have been fortunate to have many thorough, critical, and constructive reviews.

Motivated only by charity, Ruby Larive provided field assistance in my photographic efforts, and Margie Rose constructed the first draft of the text's diagrams with the aid of her computer. I also am indebted to my neighbor, the Reverend Armand Larive, for his advice on philosophical and theological points, and to my mother, a cheerful and patient proofreader.

TO THE STUDENT

Science is the most successful part of our common intellectual life. But scientific progress does not come about because robotlike researchers in white lab coats objectively investigate neutral questions. This book discusses how science really works, with special reference to the scientific study of the Earth—the discipline of geology. The question of why science works as well as it does, and where the background assumptions of scientific work originally came from, will introduce you to surprising aspects of the life of the mind.

This text reviews some of the most fundamental debates in geological science. Although not meant as an introduction to every subfield in geology today, it will acquaint you with some of the basic concepts and terminology of Earth science. One of the wonderful aspects of studying geology in any form is that you will see what you have studied around you for the rest of your life. Rocks, mountains, and fossils can be enriching parts of your business, travel, recreational, and personal lives. Further, the debates concerning the fossil record—and what should be taught about it in public schools—promise to continue into the next several decades. What you were taught in high school biology or Earth science, and what your children will be taught in such classes, depends in large measure on the wrenching social debate that this book considers in its final pages.

CHAPTER 1

Introduction

This small volume has a broad and ambitious task. It addresses the question: What is the scope and nature of modern, scientific work? Such a question is wide ranging, and all answers to it are necessarily interdisciplinary. The question cannot be answered only in abstract terms, since science exists in the culture that has shaped it and that it, in turn, has substantially changed. As you may imagine, individual scientists approach this question quite differently, and therefore this book is only one segment of an ongoing discussion.

This volume is one small effort toward educating those who will make the decisions that will affect our only home, the Earth. It is meant to be a guide for a tour of a particular branch of science, namely, geology. In addition, it introduces some basic ideas from the philosophy of science. Topics and ideas from the history of science, and also from religion, are necessary detours along the way.

Science has been linked to the most dramatic aspects of recent human history. For example, the technology spawned by science has meant that for the first time in the history of the world, hundreds of millions of people lead lives of material abundance in a sophisticated, increasingly international culture. Those of us from industrially developed nations have explored distant sectors of the earth and brought back the scarce natural resources demanded by modern industries. We thus mine ore in tropical jungles and drill for oil in Arctic plains. Indeed, we have even begun exploring beyond the earth: we have sent astronauts to the moon and complex probes to other planets.

But the scientific culture that has made these remarkable advances possible also has proved, at least to date, ill-equipped to address the problems associated with technological development. We struggle to see how we can cope with increasingly deadly weapons, with abundant and noxious pollution, with the many disparities between life in the First and Third Worlds, and with a global population that now strains the entire biosphere.

Can scientific research continue to give us the advantages that technology has brought while saving us from the problems of industrial society? To understand the possible answers to that question, we must decide what science actually can accomplish. First, science, or the scientific method, can address only those questions to which it has been adapted. Second, we must understand that science possesses both strengths and limitations.

The questions this book will examine include:

- What is scientific research?
- Is geology a science even if many of its claims cannot be tested experimentally?
- Are scientists objective and impartial? Do they inject their personal values into their work in ways that affect their findings?
- Can scientific research legitimately comment on morality, values, or the meaning of life?
- What do geology and biology say about the origin of life on our planet?
- Must science and religion always conflict?
- How does "creation science" differ from geological interpretations of the history of life? What, if anything, should be taught in our public schools about life's long and complex history?

We will begin our tour of these issues by looking at the methods that all scientists use in their research. Those methods, we will discover, are more complex and more human than nonscientists often assume. The investigations that we will look at here include some of the debates in both earlier geologic questions and current geologic research. For example, early scientific thinkers disputed the fundamental nature of certain earth materials. Geologists struggled with this question in an attempt to understand how fossils and other curious stones were formed. (The biological origin of fossils was far from immediately obvious.) We also will see how physical geologists have attempted to understand the origin of rocks and mountain chains. The origin of mountains was particularly puzzling until the advent of modern geological theories in the 1960s.

Next we will recount the early history of the earth and scientists' speculations about the origin of life. This topic will bring us close to the limits of what science can explain. However life originated, the fossil record testifies that it has had a long and rich history on this planet. Geology describes both the great span of geologic time measured on the earth and the complex changes in life-forms that inhabited the earth in the past. Evolution and extinction continue to be challenging areas of scientific research. Paleontologists (geologists who study fossils) differ with one another about important aspects of evolutionary theory and also about the nature and causes of major extinctions. There is, however, one point on which we all agree: modern humans, known scientifically as *Homo sapiens sapiens*, appear to be the first species that can consciously choose to eliminate other species from the world. The self-destructive power of this idea becomes apparent when we consider how little we know about the interdependence of organisms and the causes of mass extinctions.

The topics that geologists investigate may seem numerous and disparate, but surprisingly, all the scientific debates we will consider center on one simple question: how rapid is the rate of change in the earth's history? Many geologists and biologists believe that the history of the earth has proceeded very slowly and have based their research on this assumption. But recently geologists have decided that the changes on our planet can be as rapid as the impact of a meteorite or as slow as the evolution of blue-green algae. The current flexibility in accepting a wide range of rates of change makes geologists better able to interpret creatively the complex evidence of the geologic record.

To bring our tour to a close, we will review the main themes of this book by surveying the debate about "creation science" that has so strongly affected science curricula in our public schools. Our final question will be, Should an introduction to scientific reasoning and the evidence of the fossil record be required of students in public schools? Let's begin our work by considering a real-world geological project.

An Actual Project

We all have read dramatic announcements by scientists in the newspaper. But we don't often hear about what scientists do on an ordinary day—that is, a day when they aren't issuing dire warnings about the greenhouse effect or meteorites en route to obliterating the earth. Of course, I can't speak for all scientists, or even all geologists, but I can describe my daily routine for a project that I worked on for three years. Although this project was not especially important, it is a good example of what geologists actually do with their time and how the practical world, including funding constraints and cooperation with colleagues, often shapes the direction of scientific research.

The Background

In the mid-nineteenth century, after fighting a war with Mexico, the United States acquired what is now the American Southwest and California. Shortly before sovereignty was transferred, a sawmill operator in California discovered gold in the bottom of a stream near the Sierra Nevada Mountains. The discovery launched a gold rush unmatched by any other in American history. Tens of thousands of fortune seekers poured into California in 1849 looking for gold. Native tribes were pushed aside, and some natives killed outright, by the incoming flood of prospectors. A few forty-niners were rich before the year was over, but many searched futilely in the gravels at the bottom of rushing, snow-fed creeks.

The men who arrived in California too late to stake a claim in the streambeds began looking for "lode" (vein) deposits in the solid rock of the Sierra Nevada. Discoveries of this type of ore followed quickly. The prospectors then went underground, using picks and dynamite to mine the gold-rich quartz veins. Many saw this search as a quest for the *mother lode,* the ultimate origin of precious metals deep within the earth.

In those days, gold ore was processed and refined using liquid mercury. Mercury, therefore, was in great demand. Prospectors in California's Coast Range Mountains, near the Pacific Ocean, discovered mercury sulfide deposits in crumbly, altered rocks. Active hot springs belching sulfur-rich gases dotted the area containing the mercury-rich ore. Hundreds of men working in the Coast Ranges quickly extracted the mercury for use in the Sierra Nevada mines. In at least one hot-spring mercury mine, the ore also contained enough gold and silver to be recovered principally for those two metals.

Following the usual pattern of mining's booms and busts, the Coast Range mines collapsed when the gold rush subsided and the demand for mercury dropped to more normal levels. After the 1860s, the hot springs were thus of interest to only those few people seeking health cures from the strange and smelly waters. The idea that some of the mercury deposits contained minor precious metals was all but lost.

The Forty-Niners' Mining Techniques

Most precious metals in the last century were mined without the help of professional geologists. Rather, the prospectors were a motley mix of farmers, lumberjacks, laborers, schoolteachers, and anyone else who thought he could strike it rich by scouring streambeds and rock outcrops for gold.

But the rich, near-surface ore that fed the gold rushes of the past has long been exhausted. In this century, mining has had to become more technologically sophisticated in order to extract, at a profit, much poorer ore. Sometimes this ore lies deep within the earth, with few, if any, surface clues to its location. In short, the forty-niners' methods are not adequate to meet current challenges.

Accordingly, geologists have stepped in to fill the demand for scientifically informed exploration in both the metals and oil industries. Professional geologists use exploration techniques ranging from sensitive measurements of gravity fluctuations, which may indicate dense ore at depth, to airborne surveys of tiny deviations in the earth's magnetic signals, which may point to magnetically conductive ore deposits beneath the surface. In the oil industry, *seismic* or sound-wave data are used to detect rock formations at depth. Million-dollar computers digest tens of thousands of seismic readings, thereby helping geologists find subterranean pools of oil. In short, the search for material riches has become a highly organized and sophisticated effort coupled with a better understanding of the earth's structure and formation.

In the past decade, geologists working for mining companies and their counterparts in academic life began to reexamine the mercury-gold ore found near hot springs in California. The academic geologists asked how and why particular types of gold ore are formed within the earth. The industry

(A)

FIGURE 1-1: (A) A sinter—a mound built up from the precipitate of hot waters (see person for scale). The thermal spring is spilling over from the top of the sinter, with water causing the dark areas. (B) At the source of a sulfur-rich, gold-bearing hot spring. The light-colored material is sulfur, and the dark material is gold-rich precipitates.

geologists drilled into old deposits to find out how much ore remained underground and how rich it might be. Science and industry thus joined forces, as they often do, as both attempted to unravel the mysteries of gold concentrations in the earth.

(B)

The Project

SUMMERTIME IN THE DESERT

Standing beside a 175° F, sulfur-belching hot spring on a summer's day with air temperatures of 105° F was more than a little wilting. I had come to California to discover how and why gold was concentrated in the rocks around the Coast Range's hot springs. Working with a series of student field assistants, I spent part of three summers taking samples of rock, soil, hot-spring water, and even the gas bubbles in the springs (Figure 1-1). My assistants and I measured several variables in the field, such as temperature and pH (a chemical parameter) of the hot springs, and filled our notebooks with observations and descriptions (Figure 1-2). We collected many pounds of black, fine-grained precipitate from the bottom of the hot springs and sampled the yellow crust of pure sulfur crystals that collected around the edges of the pools.

The geological mecca of my field area was the McLaughlin open-pit gold mine, named for a geologist from the university at which I was enrolled for graduate work. The gentleman in question had risen in the world of corporate mining earlier in the twentieth century. I visited the mine and, between the explosions that the miners used to loosen the bedrock, took samples of material across the face of the mining pit. I walked, and occasionally ran, to get out of the way of the caravans of giant, specially built trucks that carried away the blasted ore.

The gold- and mercury-bearing ore was interlaced by millions of tiny veins, all leading upward to what had been an area of hot springs in the recent geologic past (Figure 1-3). The samples were pretty—sometimes even gorgeous—pieces of chalcedony and quartz, both minerals made of what geologists call *silica*. Before the gold was discovered beneath the surface, pieces of the uppermost section of the ore had been used locally as decorative stone. The area was known to be rich in mercury sulfide, not necessarily the most healthful of materials, but local residents were in the habit of discounting the problems of mercury contamination.

Gold was visible to the naked eye in only a few samples of ore. In general, gold particles in what geologists call the "hot-spring environment" are micro-

FIGURE 1-2: Multimeter used for pH measurements in the field. Certain chemical measurements at the hot-spring site are crucial to a scientific understanding of how the gold is transported in these waters.

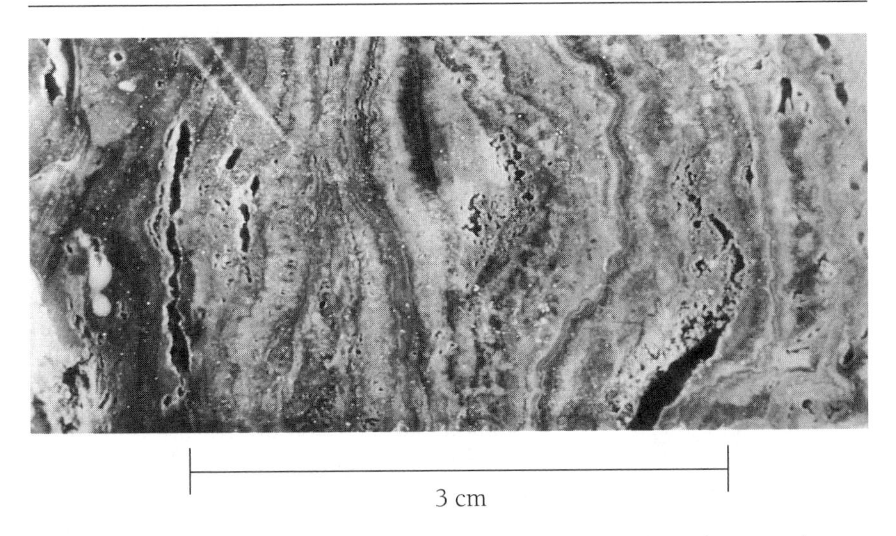

FIGURE 1-3: Small veinlets of chalcedony, a mercury- and gold-rich ore from near the surface of the McLaughlin gold deposit and mine. The gold in this sample is microscopic. The dark bands are cinnabar, a mercury-rich mineral. Voids, or empty spaces, appear black.

scopic. This has led to the name *invisible gold* for many recently discovered deposits of this type. In the field, no one can see the gold, an obvious challenge for prospecting. I was happy to collect numerous samples of the ore, confident that the gold in them would be visible under the microscope in the laboratory.

Rattlesnakes were ever present companions in the field, both around and between the hot springs and the mine. I had grown up with some experience of rattlesnakes and in fact, when I was younger, had taken pride in dispatching them to the next world. Two of my field assistants, however, had never seen rattlers before, and they worried about their presence. The snakes certainly helped keep us all alert.

In the rural counties where we worked, the days ran together in one long blaze of searing sunlight and manual labor. Our daily existence was punctuated only by retreats to air-conditioned motel rooms and stops for terrible coffee in small-town cafes. Since we covered many miles of territory with only primitive roads, we were dependent on four-wheel drive vehicles for transportation. The company that owned the open-pit mine supplied us with four-wheel drive pickup trucks. In this and other regards, the company's cooperation was based on the hope that my academic research into the chemistry of the hot springs might bring to light something of practical value for exploration geologists. Although I was grateful to have the trucks, over the three summers we used them, my assistants and I experienced breakdowns ranging from leaking steering fluid to shredded belts to the failure of a backup gas tank to feed gas into the engine. Like most geologists in the field, we also had infrequent but unpleasant episodes of dressing downs from angry property owners who objected to our tendency to trespass. More rarely, we had encounters with suspicious old-time prospectors who still worked tiny ore deposits in the more remote areas of the hot-spring district.

In short, our summer workdays, far from being easy, required more than a bit of dedication to survive. To top it all off, there was always, always, always the intense heat. Usually the daily high temperatures were above 100°, and ripples of heat drifted up from the rocks around us. Under these conditions, the sulfur gases of the hot springs were particularly sickening. Adapting as best we could to the climate, we began working at dawn and retreated toward town about noon. Even with this regimen, I was overwhelmed one day by cumulative dehydration and lay down in the dirt near a hot spring. I felt so awful that death looked like it would be a blessing. But someone else was with me and forced me to work some water into my overheated system. So, the next day, the work continued.

How This Project Started

One of the lessons of this book will be that scientific research always takes place in particular historical and personal contexts. I had become interested as an undergraduate in problems associated with nonrenewable resources because I thought they were becoming increasingly important to our country's economy. Politics was, and is, a special interest of mine, and I believe that political possibilities are shaped, at least in part, by economic realities.

As an undergraduate I had concluded that academic economists did not understand the direct, physical constraints imposed on the oil and metals industries. Similarly, economists did not have the background in chemistry necessary to understand what was, and was not, possible in the areas of pollution management and control. No doubt the arrogance of youth made my judgments too simplistic, but nevertheless they propelled me into studying geology and applied chemistry. So, because of a passion for politics, I became a geologist and geochemist.

My specific interest in the gold-mercury deposits of California was fostered by my friendship with a colleague (a fellow graduate student) who was working in the same field area. He was an outgoing and talkative man, and because we happened to share an office at the university, I heard more about his work than about anything else happening in our department. My office mate's interest in the gold/mercury subject had been encouraged by our mutual adviser, a senior professor who valued cooperation with the company that operated the open-pit mine in the area. Thus, because of a professional friendship, on the one hand, and the values of an academic adviser, on the other, I took up my studies in the Coast Range as part of the requirements for my doctorate in geology.

The Problem of Ownership: Mineral Rights and Access to Mines

Interest in a given geologic problem is only a starting point. Since most of the earth is divided up into parcels of private property and since all known gold deposits are guarded by the mining companies that own them, academic geologists must have productive ties to the people who control access to geologically interesting areas. In the case of the hot-springs ore in California, this meant cultivating and maintaining a professional relationship with the geologists and administrators of the mining company operating the open-pit mine and exploring some of the nearby hot-springs property.

Gold-mining companies are notoriously paranoid about their operations. Exactly where they think gold might be located and how much gold may still remain in operating mines are closely guarded secrets. A person without professional connections in the industry will get no reply to questions about exploration strategies or ongoing projects.

The university at which I was enrolled had a long-standing relationship with the American mining company operating in California. Both the company's former president and the then current chief of exploration had graduated from my university, and my adviser had worked hard to maintain a productive relationship with the company. Indeed, it was because of these connections that my office mate had begun his work in California a year before I entered graduate school.

These continuing professional ties to a particular company gave me access to the open-pit gold mine. Without that access, I still could have worked on the nearby hot springs, but the results would have had considerably less

FIGURE 1-4: Although beautiful, these mountain peaks in northeastern Oregon are accessible only during a few weeks of summer, and then only after considerable personal effort. Geologists find it easier—and cheaper—to conduct research at lower elevations.

geologic meaning. Sometimes geologic work in foreign countries or on native (tribal) lands presents another set of problems. Accordingly, some parts of the globe have been studied thoroughly more as a matter of convenience than anything else. It should come as no surprise that difficult environments, like the highest elevations of the Andes or the tundra of northern Canada, have not been the subject of as much geologic research as, say, the rocks in New York State. Again, practical matters sometimes outweigh the purest of scientific agendas.

The Problem of Funding: Public and Private Money

Access to samples is not enough. Travel and lodging, laboratory expenses, computer time, and researchers' salaries all must be covered in order for any research project in geology to progress. I was fortunate to have an academic fellowship from the National Science Foundation which paid me a stipend for several years. But to work on the hot-springs project, I clearly needed other sources to pay for expensive laboratory work and repeated visits to the field.

The mining company to which my adviser had ties was the source of small grants and help "in kind," for example, pickup trucks in the field. But for the scope of work I envisioned, I knew I would need a great deal more. Because my academic adviser was not used to thinking in such terms, I turned to my undergraduate adviser for assistance. This professor had a long track record of obtaining large grants from the biggest source of funding in geology: the Earth Science Division of the National Science Foundation (NSF).

Over the telephone, my undergraduate adviser described the impression that a successful grant proposal must create, and he gave me tips about writing it. His advice drastically changed what my office mate and I had previously envisioned for our cooperative grant proposal. With this new information, we wrote a proposal structured as my undergraduate adviser recommended. Again, particular professional connections were crucial.

A professor at another New England university had developed a special technique for occurrences like the hot springs, based on chemical measurements made possible by, of all things, a particle accelerator. That professor prepared a separate NSF grant proposal in parallel with our project. In the end, both proposals were chosen for funding. In total, the work with which I was involved consumed about $300,000 of taxpayers' money, much of it absorbed directly into the universities' overhead charge or "tax" on the project.

Our success in obtaining funds was unusual, as most grant proposals in the geological sciences received by the NSF are rejected. The reason is not that the proposed projects do not represent good science or that the scientists involved are not qualified. Rather, the recent reductions in the money Congress makes available for scientific research mean that most NSF proposals cannot be funded. Therefore, the competition for research money is a major force in shaping what research is actually carried out in geology today. This may not be what geologists would choose in the abstract, but it is dictated by increasing laboratory costs and decreasing funding for scientific research.

I also should note that some geologic subjects have been more lavishly funded than others, by both public and private sources. Work that might be useful to the oil industry, for example, has been given more financial support by various granting agencies than, say, work concerned with the earliest fossils of primitive life-forms. Some scientific projects are simply judged more useful to our society than others.

Working in Laboratories

One aspect of research in geology that may be surprising is the degree to which scientists depend on a network of different laboratories. Specific techniques are often available at only one, or relatively few, labs in the United States. Working in different laboratories, and therefore cooperating with a number of scientists and technicians, is a normal part of research. The social nature of science in this respect is clear from the example of my hot-springs project. Before all the work was done, many laboratories had contributed their expertise or allowed me to work in their facilities. The list included:

- two government laboratories (both part of the U.S. Geological Survey)
- five universities with their assorted lab facilities
- two commercial laboratories (one specializing in gold assays of rocks).

The complex infrastructure represented by these laboratories is part of a societywide investment in scientific research made in this country after World War II. The cost of these facilities is staggering, but many scientists and politicians alike would argue that if the United States did not have such laboratories, we could not compete in the global economy.

The cooperative and social characteristics of most research in science today mean that only certain sorts of people excel at the work; most "loners" soon drop out. The image of the isolated, "mad" scientist is thus inaccurate.

Furthermore, the social nature of science means that the research is marked by an inherent conservatism from beginning to end. That is, strange new ideas are resisted—not necessarily rejected out of hand—but resisted. As we shall see later in this book, fundamentally new ideas have, in fact, emerged in twentieth-century geological research. But it also is true that the cooperation that scientists need generally means that what most geologists perceive as "weird" ideas or "wacky" research projects never materialize. This conservatism inherent in science is not much different from the inertia present in all professional fields. No doubt it often saves geologists from considering unproductive ideas, although it must occasionally inhibit unexpected advances.

Making Computer Calculations and Interpreting Data

After making laboratory measurements in facilities from Massachusetts to California, the next step of my project was to interpret the results—the data—I had gathered. Some points were instantly clear: ore-grade gold was present in the precipitate of the hot springs I had studied. This news, exciting though it was to mining companies, was also quite unexpected to the academics with whom I worked. It opened up a challenging chemical question: how could waters like the hot springs dissolve and transport gold to the very surface of the earth?

Gold is notoriously inert, or unreactive, which is why we use gold in dental work and wear gold rings on our fingers. We value the fact that it doesn't react with its environment; that is, it doesn't tarnish or rust, much less dissolve into nothing. Geologists were used to thinking that processes in the earth that could keep gold in solution were exceedingly rare, that they could occur only at much, much higher temperatures than what I had seen in California. Also, waters capable of dissolving gold were expected to be extremely salty, much more saline than my samples. Clearly, the hot springs in the Coast Range Mountains were producing something both delightful and unexpected.

Next, I worked in the laboratories of my university, studying the gold ore from the open-pit mine. Using microscopic and physical determinations, I concluded that the waters that had long ago flowed through the ore deposit and created the beautiful chalcedony and quartz veins were quite similar to the waters still flowing in the active hot springs. Clearly, I was on the trail of a process that had real geological significance; the hot springs were much more than a mere curiosity.

Other parts of the laboratory results addressed more difficult questions that need not be described here. But in the end, one of the most interesting questions was, How did the gold stay dissolved in the hot-spring waters? Why did it precipitate in the hot springs rather than flowing downstream with the rest of the water?

Computer calculations based on the branch of chemistry known as thermodynamics provided an answer. The high sulfur content of the hot springs, the calculations indicated, was the key to keeping the gold in solution. When the sulfur left the hot-spring pools (in the form of the gas that smelled so strongly and the yellow crystals of sulfur at the pools' edges), the gold precipitated out of solution onto the bottom of the hot-spring pools. Similar waters had, in like manner, formed the ore being blasted out of the open-pit mine I had studied. This result was just different enough from the expectations of many geologists to be interesting to all who saw the data. The ideas I generated were not radical, but they provided a new twist on the old problem of how earth processes could lead to high concentrations of gold in a few localities.

Publication and Response

Because part of any successful scientific work is communicating the results to others, my colleagues and I wrote up the data and calculations from the hot-spring studies in a series of five abstracts (very short papers), presented to the geological community at national meetings, and four full-length articles published in major scientific journals.

At each stage in the process of publication, editors, reviewers, and coauthors helped shape and improve the reported results and refine interpretation of the evidence. Again, science progressed because of cooperative work among many different people. After our work was published, it was gratifying to receive requests for more information from geologists around the world interested in specific parts of the California study.

The Rewards

As expected, financial rewards from the project have been nil. That's right, nothing! Mining companies are glad to use academics' work, but they do not pay for the results, which are a matter of public record. In this case, federal tax money from NSF subsidizes industry research in geology. But, although I was never paid for my results, I did earn a respected credential in the scientific community, namely, my Ph.D.

Like most of my classmates in graduate school, I did not find a job quickly after leaving school. American society, for better or worse, is employing fewer and fewer of the scientists that our universities produce. Most of my classmates in the sciences have left the field altogether, and two have found research work only in Germany. After several years' effort, I was fortunate enough to find a teaching position in the geology department at the state university in my hometown.

Unlike the "real-world" consequences, the intellectual rewards of the work were great. It may be difficult to imagine, but some of my finest moments came while I was studying complex mathematical problems and performing thermodynamic calculations. The work literally led to tears of joy. The possibility of understanding a process that had previously been opaque to science is what really drives, and rewards, researchers everywhere.

The Cycle Begins Again

Many, indeed most, questions about "hot-spring" or "invisible" gold deposits remain unanswered. Other geologists have taken up where our studies ended. The cycle of generating ideas, searching for funding, cooperating with mining companies, and trying to interpret complex results has begun once again. Scientific work, in this regard, never ends. Even a good, convincing answer to one question leads to other ideas and speculation.

SUMMARY AND CONCLUSIONS

Science and technology form the basis of our industrial society. But whereas some people look to science to solve problems, others distrust scientists and the culture in which they work. This book offers a framework for considering what science is and what it can do. We examine specific cases of geological research as we explore scientific methods and assumptions.

Just one, simple example of research into the formation of "invisible gold" deposits makes it clear that modern science is:

- relatively expensive (and usually publicly funded)
- highly social (few "loners" survive long in science)
- professionally conservative (as are all disciplines)
- demanding but intellectually rewarding (especially for those enthralled by the natural world).

CHAPTER 2

The Scientific Method: An Empirical but Deeply Human Undertaking

Many people believe that scientists use special methods that place scientific work on a plane different from other human endeavors. And indeed, scientists' methods are distinctive. Scientific work reflects an **empirical** approach to learning about the world that is based on the facts of experience and a particular style of reasoning. One feature of empirical work is that the evidence that researchers use is public (or external) rather than private (or internal). This means that other people can verify the facts that scientists use. In contrast, other areas of intellectual life are usually considered to lie outside the empirical realm. Mathematics, music, and sports, for example, are valuable parts of our lives that are not part of the empirical world of science.

In modern Western history, the tensions between science and religion have been severe and have most frequently been resolved in favor of science. In large measure, we live today in a society dominated by empirical reasoning, but we shall see that the method of science—that is, empirical methodology—is not as crisply clear as our society often assumes. Instead, scientific research is a deeply human undertaking, marked by the foibles and virtues of those who devote their lives to empirical knowledge. Nevertheless, at its core, science is concerned with the world external to the scientist, resting on the assumption of a simple cause-and-effect relationship among natural (external) phenomena.

The History of Scientific Thought

THE BEGINNINGS

Humans existed for millennia without science. During this time the world was generally regarded as the stage for supernatural forces. Myths and religious ideas explained the natural world. Earthquakes and floods, for example, were often viewed as expressions of divine wrath in response to human transgressions.

Another ancient approach to thinking about the world was to assume that physical objects had many of the same characteristics as persons. Mountains, for example, were sometimes thought to have souls, and roaring fires were believed to have intentions and desires. Although we may laugh now at such ideas, our language still reflects them in such phrases as:

Nature *abhors* a vacuum.

Water *seeks* its own level.

Heat *escapes* through the kettle's spout.

Opposite charges are *attracted* to each other.

In prehistoric times, both supernatural and personal explanations of events in the physical world held sway. Although the history of humanity moved slowly during this time—a period spanning most of the existence of *Homo sapiens sapiens*—the scene changed in the city of Athens during the Classical period of Western history.

Several Greek theorists, most notably Aristotle (384–322 B.C.E.), offered a different style of explanation for the events in the natural world. Aristotle taught that explanations of the natural world should be limited to those that could be verified by the facts of experience. Aristotle and his followers tried to describe what they observed in terms of regularities embedded in the natural, not the supernatural, realm, and this was the foundation of scientific work. Aristotle's style of thought was incorporated into Greek astronomy, which much later became a model for the other sciences to emulate.

THE NEXT STEP

Although the Romans knew about the Greeks' intellectual achievements, they were more interested in ruling an empire than in developing abstract thought. As the Roman Empire waned and Christianity waxed, St. Augustine (354–430 C.E.) complemented Aristotle's groundbreaking work by inventing a field of his own, the discipline we now call *theology,* the systematic study of God and God's relationship to the natural world. The world, Augustine believed, was best explained as God's handiwork, which meant that natural phenomena should be understood with reference to spiritual life.

Augustine's work was not empirical. Following the ideas of several early church fathers, he believed that the truth was revealed by God in Scripture, not constructed by the observations and reasoning of Greeks. In fact, the truth might contradict reason, simply to test a person's faith.

During the period immediately after the fall of Rome, intellectuals in Europe dismissed the Greeks' accomplishments and instead based their thinking on Holy Scripture, complemented by appeals to St. Augustine's authority. Fortunately, from a scientist's perspective, Islamic scribes and theorists preserved the spirit of Aristotelian thought and copied Arabic translations of the Greek works.

As the Middle Ages blossomed, the Arabic versions of Aristotle were translated into Latin, and Christian theologians came to accept many ideas from the classical world. Aristotle thus became an authority like Augustine, to whom an educated person could appeal in an intellectual disagreement. Although the works of the Greek writers were studied throughout Christendom during the late Middle Ages, the institutional church, not surprisingly, still valued a supernatural framework for explanations of the natural world.

For Augustine and the early church, a statement was true if it accorded with God's word or, more fundamentally, with the will of God. True statements had value because they agreed with what was divine. There could be no distinction between the truth of the **facts** of the world (for example, the existence of *Homo sapiens sapiens*) and the **values** a person rightfully invested in them (we humans have value because we were created by God and are part of his plan for the world).

For modern readers, the refusal to distinguish between "facts" and "values" is often quite surprising. Indeed, it is almost unintelligible to some members of our society. Many people assume that facts are objective and actual parts of the world, whereas values are subjective and personal. But such a view is not only foreign to much of the religious realm, it is also rejected by some modern philosophers (see Appendix A).

The Crux of the Disagreement

Early scientists often were at odds with the established church because science had a wholly different way of approaching facts. Science assumes what philosophers call **efficient causes**, that is, causality linked to an immediately preceding event that has a direct effect. Because of professional habit, scientists always explain one phenomenon in terms of another, operating in the recent past. For example, if your roommate gets angry and throws a book on the floor, you will hear a loud thump and may jump from your chair. Or because gold particles are denser than other geologic materials, gold nuggets accumulate in the bottom of a stream while other, less dense, particles are washed away.

Modern science leaves out another type of causality, that of **final causes**. This causality is dependent on broader themes and on the future, not the past. Final causes are assumed by most religious thought and some secular philosophies as well. An undisputed example of a final cause could be your plans about what you need to accomplish this week: your intentions might lead you to undertake a number of different actions (preparing for a test, writing to your sister, practicing the fluegelhorn, filing your tax return) not linked to one another by any efficient causation. Similarly, when I pack my suitcase and set my alarm because I plan to catch an airplane in the morning, I am illustrating final causality. But it still is proper that I, as a scientist, look only for efficient causes in my laboratory data, for that is the task that modern science has set for itself.

Contrary to most scientists today, Aristotle subscribed to final causes. He believed, for example, in the ongoing spontaneous generation of life, and that in the inorganic world such generative powers indicated a latent force striving to manufacture life. Generating life almost *ex nihilo* (from nothing) was an indication that life-forms were the goal of the world, the final cause of the world's existence toward which the natural world was moving. In early modern Europe, however, final causes were championed by theologians, not the intellectual inheritors of Aristotle's basic empirical approach to knowledge.

SCIENCE VERSUS RELIGION IN THE MODERN ERA

The first of many battles between those subscribing to efficient causality (the modern discipline of science) and those subscribing to final causality (often represented by the Christian Church) is deservedly famous. Galileo Galilei (1564–1642), an Italian inventor and scientist of the Renaissance, defended the idea that not all heavenly bodies revolve around the earth. He did so because of his empirical observations of the heavens. The church, however, taught that the earth must be at the center of the universe because Scripture emphasized the divine creation of the planet and divine interaction with humankind. That is, solely on biblical grounds (as interpreted by Catholic scholars), the church stated that the facts of astronomy contradicted what Galileo was arguing. Then, because the church threatened Galileo with torture for his heretical findings about the natural world, one of the greatest thinkers of the Renaissance was effectively, if temporarily, silenced.

From our twentieth-century perspective, it is easy to make the church the villain of the story and Galileo, the hero. Scientists, of course, are comfortable with such a reading, as most believe that neither philosophers nor theologians should meddle in research into the natural world. But it is important to remember that the same institution that persecuted Galileo also preserved Greek ideas through many difficult centuries, doing so because such valuable knowledge revealed God's intentions. The church, therefore, was acting consistently when it censored Galileo, asserting that the "facts" of the natural world had to square with God's plan. Finally, and ironically, Galileo's career was facilitated by an institution the church had created: the university.

Furthermore, because the scientific revolution developed in Western culture and nowhere else around the world, it is natural to speculate about how Western religious thought affected the intellectual foundation for science. The Aristotelian heritage is clearly one element in the equation. But it may be that the Judeo-Christian tradition was equally important. As we shall see, the basic assumptions of Western science were developed by Christian theologians as part of their view of one, all-powerful and all-knowing, creator God. But, early Western scientists discovered, such thinking could also be used in strictly secular explanations for natural events. The ensuing tensions between early modern science and the church increased with each decade, despite some fundamental points of agreement.

OUR CURRENT ATTITUDES

Today, like Galileo, most of us assume a secular, scientific framework of thought in regard to the world around us. When someone asks, "What causes the common cold?" or "Why did Los Angeles have another earthquake?" that person is usually looking for answers that appeal to the regularities of the natural world that we all can observe. Similarly, if an elderly man suffers intermittent chest pains and goes to a doctor, he expects the doctor to perform tests and look for efficient causes of the pain. We would think it absurd if the doctor explained the pains with reference to supernatural forces or a final cause.

> Those of us living in the modern world want a system of analysis that will allow us to go further and further toward understanding natural phenomena. We have put an enormous amount of trust in Aristotle's hopes for explaining the world on an empirical and naturalistic basis.

For better or for worse, we live in a technological age, and we have surrounded ourselves with a culture shaped by scientific thought. Even the religiously devout almost always want to be taken to a doctor, not an altar, when they have a heart attack. This is hardly surprising when we remember that explaining a phenomenon by referring it to God allows us to explain everything in a similar manner. For example,

God decided to give Los Angeles an earthquake this morning.

God has sent the HIV virus as a punishment for sexual transgressions.

God made my sister rich but made me good looking.

These explanations are logically acceptable as far as they go, but they cannot go any further. Speculating about why God chose this, rather than that, is beyond us.

Scientists indirectly encourage other people in our society to trust them when they claim credit for a dazzling array of technology. In fact, we could say that science now functions as the church once did, as the external, legitimizing institution for our society. As Harvard scientist R. C. Lewontin wrote, "Science claims a method that is objective and non-political, true for all time. Scientists truly believe that except for the unwarranted intrusion of politicians, science is above the social fray . . . the product of science is claimed to be a kind of universal truth."[1]

Scientists also like to lecture those outside their discipline about the difficulty of their work and the rigors of scientific methodology. A powerful

[1]R. C. Lewontin, *The Doctrine of DNA: Biology as Ideology* (Harmondsworth: Penguin Books, 1991), p. 8.

methodology, it is often assumed, is what makes scientific work so valuable. Accordingly, scientists often try to reinforce the belief that only the elite can properly understand scientific findings. Thus we have reached the point at which everyone in our culture is deeply affected by science, but few people seriously study the subject.

The Scientific Method(s)

WHAT IS THE SCIENTIFIC METHOD (OR METHODS)?

To the surprise of those outside the field, there is scant agreement among scientists about what constitutes the scientific method. Indeed, opinions about this matter not only differ; some of them are contradictory. Many scientists hate to examine questions about their methods, although they all are sure their own work qualifies as "science." Such disinterest in definitions and in critically considering methodological questions may seem narrow-minded at first, but perhaps it is not wholly unreasonable. Most poets, after all, do not spend time defining the nature of verse but instead put their energies into writing poems.

Traditionally, science was considered an art, one of the liberal arts used to train young minds in colleges and universities. According to this perspective, science has more in common with literature than it does with engineering. But unlike the humanities, science claims that it is less directly concerned with *Homo sapiens sapiens* than it is with the nonhuman universe around us. Science, we sense, is engaged with the outer world in a way that the humanities appear not to be.[2]

THE TRADITIONAL CLAIM

The traditional view of the scientific method is that science is **objective** (that is, it is known because of the world outside the mind). It also holds that science is impersonal and timeless. Successful work that follows the scientific method, it is believed, leads to greater and more fundamental truths. Science rests on experiments that can be reproduced, a feature of science that distinguishes it from all the other arts of a liberal education. In addition, it has

[2]To the surprise of scientists, many philosophers dismiss the notion that there is a meaningful distinction between methodology and substance. To them, the gap between "facts" and "assumptions" narrows to nothing. This book, however, is written from the scientific perspective, however philosophically naive that may be.

been pointed out that scientific methods reach across cultural boundaries. Despite differences in race, religion, and language, one can find science being practiced in much the same way in China, India, and England.

According to this general view—still taught in many high schools across the country and even in some colleges—scientists proceed with their work according to the following method:

1. Making and recording observations.
2. Formulating a question about these observations.
3. Proposing a hypothesis to answer this question.
4. Gathering data to support or disprove the hypothesis.
5. Drawing conclusions from the data to accept or reject the hypothesis.

This view tacitly assumes that scientists are not influenced by personal beliefs, by their social class or standing, or by professional pressures from inside or outside the scientific community. Scientists themselves, it is thought, have such impersonal working lives that they are interchangeable with one another. That is, Jane and John will reach the same conclusions if they both follow the scientific method.

A REALITY CHECK

The traditional claim concerning the scientific method, widespread though it may be, can be shown to be false by a cursory reading of the history of scientific debates. Individual preference, professional standing, competition for funding, and personal jealousy all influence scientific research. Indeed, some scientists characterize their professional relationships as "cat fights," clearly a rather personal approach to their work.[3] And although scientists in China, India, and England all may follow the same fundamental rules, they often investigate entirely different questions, a choice determined by cultural considerations and individual preferences.

None of these arguments means that science is as subjective as poetry, but merely that scientific research is a thoroughly human and creative endeavor. Because scientists are people, their personal differences do indeed show up in

[3]Bernard Davis, a professor at Harvard Medical School, argued that scientific *research* may be subjective, whereas scientific *knowledge* is objective. He has been one of our society's most sophisticated defenders of scientific objectivity, and his essays stand in opposition to most published work in the philosophy and history of science.

their approach to research. Although they try to concentrate on the non-human aspects of the universe, they never fully leave themselves behind. According to Stephen Jay Gould,

> The stereotype of a fully rational and objective "scientific method" with individual scientists as logical (and interchangeable) robots is self-serving mythology. . . . [T]he messy and personal side of science should not be covered up by scientists for two major reasons. First, scientists should proudly show this human face to display their kinship with all other modes of creative human thought. . . . Second, while biases and preferences often impede understanding, these mental idiosyncrasies may also serve as powerful, if quirky and personal, guides to solutions.[4]

Scientists usually frame their work by enthusiastically **advocating** a particular viewpoint. Published scientific work is rarely characterized by doubt or uncertainty, nor does it often reflect an evenhanded evaluation of opposing theories. Instead, scientific authors present one view and all the evidence they can find in favor of it, rather like a lawyer arguing in court on behalf of a client. This tradition of advocacy can polarize researchers who may, in fact, have many ideas in common. It may not be surprising that in the heat of debate, scientists are sometimes carried by their enthusiasms into assertions and attacks stronger than the available data warrant.

To illustrate these issues, think of a continuum (Figure 2-1). On one end of the spectrum are those people who believe that scientists almost always accept or reject theories because of experimental evidence or observation. Scientists, as you might guess, are fond of the view that theories are accepted on objective grounds. Members of the public all too often assume this viewpoint without critically examining the reasons for it.

(A word about the term *theory* may be useful here. In science, a theory is not a guess or a casual thought; rather, it is a synthesis of measurement, calculation, experiment, and critical thinking. This idea can be confusing because in colloquial English, the term *theory* may mean nothing much more than an instantaneous opinion (for example, "I heard the evening news last night, and my theory about that ax murder is. . . ."). But in scientific work a theory is the rigorous organization of a great deal of knowledge about the natural world.)

On the other end of the continuum are some social scientists and historians of science who believe that social considerations can influence scientists

[4]Stephen Jay Gould, "This View of Life," *Natural History*, February 1994, p. 14.

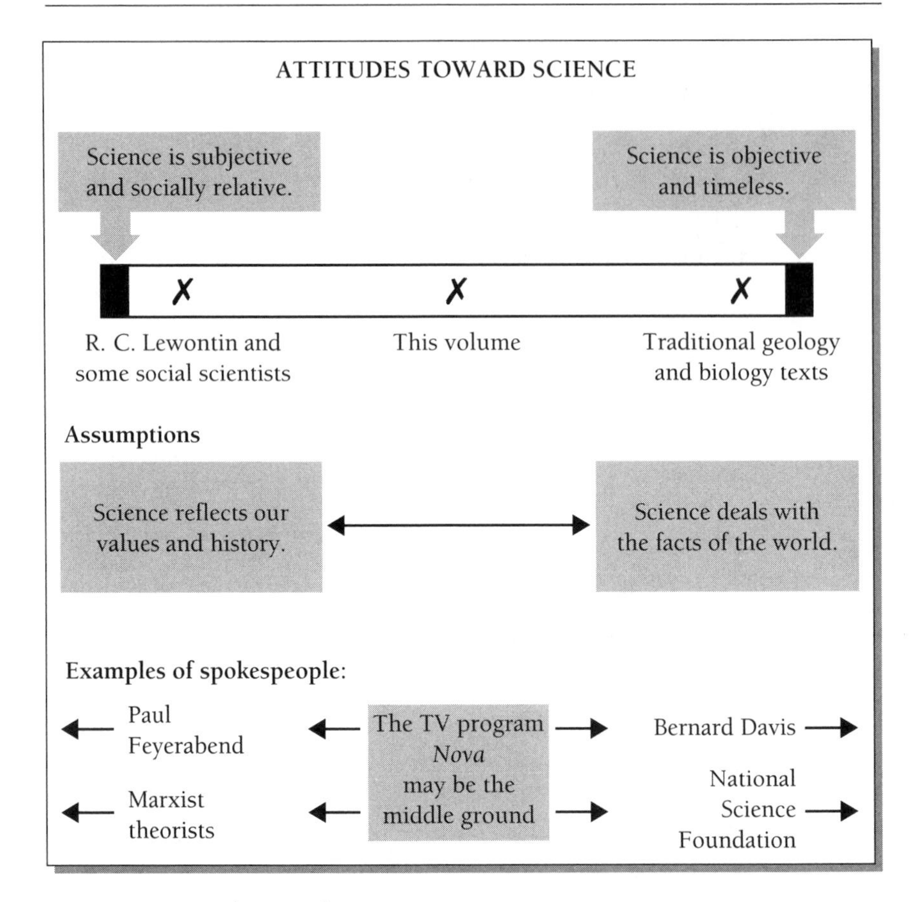

FIGURE 2-1: Attitudes toward science in our society.

just as much as laboratory data do. They point out, for instance, that studies of the capabilities of the sexes, or of different races, have been marred by irrational and pernicious prejudice. And, these critics argue, the methods of science reflect the individualistic, competitive, and elitist society that gave birth to scientific research.[5] This is hardly surprising, when we remember that science is a social creation.

[5]Scientists and nonscientists who emphasize the subjective nature of science tend to be attuned to the social and political issues of the day. To put it crudely, many subjectivists are "liberals." This tendency is by no means a requirement. Nevertheless, the many connections between radical subjectivism and progressive politics, as well as their combined affect on attitudes toward science, are well but harshly highlighted in Paul R. Gross and Norman Levitt, *Higher Superstition: The Academic Left and Its Quarrels with Science* (Baltimore: Johns Hopkins University Press, 1994).

In addition, some assumptions of science have changed from what they were in the days of Aristotle or even Newton, and there is reason to believe that science will change still more in the future. Nothing about specific scientific theories, or even some points of methodology, is timeless. And the opposition to the myth of scientific perfection has grown stronger in recent years. But most philosophers and scientists still accept the notion that scientific research does follow definite rules and that scientists offer hypotheses that, at least in principle, can be tested by other researchers.

The notion of **falsification** was fully developed by the philosopher Karl Popper (1902–) and is often invoked by scientists as the core of their research. Only if data can be gathered that could falsify a scientific claim, scientists assert, can they entertain that claim in their work. That is, if no method of testing a proposition is available, even in principle, a scientist should not include it in a theory. Thus the idea of falsification is important to many scientists. Philosophers, it might be noted, have found many problems with Popper's approach.

The Method of Multiple Working Hypotheses

At the end of the nineteenth century, T. C. Chamberlin (1843–1928), a geologist and prominent scientist, tried to respond to some of the criticisms of the scientific method by developing a specific procedure for scientists to follow. He termed his idea the **method of multiple working hypotheses**.

Chamberlin argued that a scientific researcher should consider data while keeping an open mind about several competing **hypotheses**, each of which also could explain the data. Only when evidence shows that a hypothesis is inadequate should it be discarded. According to this view, scientists investigating a problem should first formalize several possible explanations of it. As impartial professionals, they should examine all their hypotheses. They should gather data that could falsify each of the hypotheses, not simply the least favored ones, thus ensuring progress on an impersonal basis.

ANOTHER REALITY CHECK

Chamberlin's viewpoint is still uncritically accepted by many geologists and is dogmatically taught in many introductory science classes. But the crucial step in the program that Chamberlin recommended is the generation of hypotheses, and unfortunately, these do not automatically leap from the natural world into the neutral and objective mind of the scientist! Instead, it is the all-too-human scientist who must invent hypotheses regarding the external world. Naturally, scientists generate hypotheses that reflect both conscious and unconscious assumptions and that come to mind only because of particular historical circumstances. In short—and as modern philosophers love to point out—no scientist can begin work simply by collecting "facts." Rather, scientists work with certain ideas already in their minds, ideas that color what may even count as a "fact" worthy of recording in their notebooks.

It is important to recognize that scientists make observations only when they are looking into something, and this "something" has often been brought to light by special and historical circumstances. What counts as a worthwhile topic of investigation is determined—or at least is greatly influenced—by the social nature of science. The work of peers, the ideas of friends, and the availability of certain sources of funding all have an impact on deciding what issues to address. In turn, these choices control what observations and measurements a scientist will make.

There also are nonsocial factors that vary through time and are equally important to what a scientist may investigate. The basic instruments and tools that have been invented and the branches of mathematics that have been developed are crucial to what research can be undertaken at a particular time. In this sense, scientific work is always tied into broader themes of human history.

Furthermore, and contrary to idealized views of science, when data are noisy or unintelligible from a particular viewpoint, they often are simply set aside. Work on the question may well be suspended. Generally, those measurements perceived as noise are not published at all. For example, the entire, exciting field called chaos theory was dismissed for centuries across many subdisciplines of science.

The history of science also makes it clear that new theories are sometimes neglected by scientists, especially established scientists, merely because of their novelty. More than a generation ago, Bernard Barber outlined this phenomenon, and the reasons for it, in his aptly titled article, "Resistance by Scientists to Scientific Discovery."[6] Thomas Kuhn put forward similar ideas in his influential book published a year later.[7] Kuhn's model has dominated a large measure of work in the history of science. Although there may be severe shortcomings with Kuhn's thought, it is clear that scientists, ranging from the

[6]Bernard Barber, *Science* 134 (1961): 596–602.

[7]Thomas Kuhn, *The Structure of Scientific Revolutions* (Chicago: University of Chicago Press, 1962).

obscure to the famous, have occasionally attacked new theories in a manner that later appears biased or even quite ludicrous.

But worse is yet to come: A few scientists are simply fraudulent. That is, they concoct data from thin air or deliberately miscalculate statistics. Scientists like to think that such researchers are rare, but fraudulent work is discovered often enough to supply the media with a constant dribble of stories about scientific scandal. And of course, there are other scientists who are not willfully fraudulent but are, to one extent or another, incompetent.

WHAT DO SCIENTISTS REALLY DO?

A cynical response to this question might be that generally what scientists hope to find in the laboratory are data offering a new twist to a familiar theory. The new data should not contradict earlier ideas but, rather, add complexity to a familiar discussion. The research described in Chapter 1 could be said to fit this pattern. It is work of this sort that is generally rewarded by publication.

Another reply, and one that speaks to all creative endeavors, is that scientists do whatever they enjoy doing. The best science is done by those who love their work. This love rests on the drive to understand the mysterious, not just to get ahead in a profession. According to Thomas Aquinas (1224–1274), for some people the need to understand is stronger than any other human drive; it is stronger than sex; it is stronger even than the fear of death.

Less dramatically, one might speak of the curiosity natural to certain minds, a kind of playful engagement with the world. Science, after all, began as an avocation (hobby). It was not pursued in an effort to earn money, fame, or prestige. Even though science has been thoroughly professionalized in the modern age, scientists still strive for something more than larger research budgets.

This joy, then, is at the core of science, the part of the enterprise that makes unexpected discoveries possible and that occasionally triggers intellectual revolutions. Curiosity and the urge to know defy cynical analysis and make science an exciting undertaking that captures many excellent minds in each generation. Fundamentally new theories are always a possibility, and the best researchers always keep that in mind, even though scientific journals may be full of more mundane findings.

A SURPRISING NOTE

The naive view of science held by many members of our society emphasizes that science is objective. Humanistic work, it is said, is subjective (that is,

"This is the one—we want you to pray for *this* one."

dependent on personal values and the nature of the mind). But subjective values also lie at the core of scientific research. Although it is a fact not well known outside the technical community, scientists sometimes choose theories for **aesthetic** reasons. Surprisingly, "beautiful" theories, especially in physics, have proved to be the most valued and fruitful. Renowned physicists of this century have preferred "beautiful" or "elegant" equations to those that better fit the experimental data! Beauty, these physicists believe, will be vindicated in the long run, and in many instances this seems to have been the case. But neither philosophers nor scientists know why aesthetic criteria have been so successful scientifically, a point we will return to later.

ENDING WITH FEWER APOLOGIES

The tone and spirit of this chapter are, no doubt, responses to the oversimplifications common in discussions of methodology in traditional science textbooks. In trying to combat the naive rhetoric of many geologists on these topics, I may have left the impression that science is deeply flawed and hinges on matters of personal preference. But doing science, like doing other intellectual work, depends on being open to persuasion by reason and evidence even when the data contradict one's own theories or preferences. Scientists might reasonably claim to have repeatedly displayed a strong commitment to objective criticism, resulting in revisions of previously cherished ideas. Science has indeed been extraordinarily successful in predicting phenomena in the dynamic and complex natural order. Perhaps no other system of thought has been so successful in its own terms.

Even though there is some fraudulent and incompetent work in the scientific realm, such work is exposed only because scientific work is also self-correcting. Other intellectual disciplines do not display such strongly self-correcting qualities. Peer review and cross-checking of results are part of science, and by and large, they work. This facet of scientific research is often ignored by critics of technical work and undervalued even by scientists. In short, science is constrained by reality in a way that other forms of human inquiry appear not to be.

Science works! We may not understand why it works, but critics of science must bear in mind its successes at every step of their analyses—at least if such analyses are to have any worth beyond rhetoric.

SUMMARY AND CONCLUSIONS

Especially when scientists are directly competing with one another, scientific research contains subjective and personal elements. None of these aspects of scientific work, however, outweighs the fact that the main reason for such inquiry is to learn about something outside oneself. Scientists do try to evaluate theories by the criterion of how well the ideas fit what they know about the physical world.

Science is both a body of laws or theories and a method of adding to this body of knowledge, with the former often thought to be more objective and impersonal than the latter. Although we perhaps would be better served by having two separate words for the *content* versus the *activity* of science, the term *science* covers both ideas.

Scientific research assumes efficient causality for natural events and is based on observations and experiments that can be verified by others. Although remnants in our culture appeal to final causes—for example, religious traditions—almost every educated member of our society regards the methods and assumptions of science as our most valuable inheritance.

QUESTIONS FOR DEBATE AND REVIEW

1. What was Aristotle's methodological contribution to the sciences?
2. Distinguish between efficient and final causes.
3. Imagine that you are in a horrible automobile accident that completely destroys the car in which you were riding. You, however, receive only a few cuts and bruises. Looking at the heap of scrap metal that used to be a car, a close friend of yours says, "You must have been meant to survive. There just must be a reason for it." What kind of cause is your friend referring to? How would a scientist reply to your friend's assertion?
4. Describe the idea of falsification. Who championed it?
5. How does a scientific theory differ from a casual opinion about the prospects of a political candidate or a baseball team? How does a scientific theory differ from a carefully considered opinion concerning a moral question?
6. If you felt great pressure and pain in your chest and believed you were having a heart attack, you probably would want to dial 911 and be taken to a hospital. If a friend of yours, in the same situation, wanted to be taken to a priest, would you think your friend misguided? Why or why not? Would you, despite your friend's clear wishes, insist on an ambulance and the emergency room? Why or why not?
7. How do scientists sometimes incorporate aesthetics into their work? Is such thinking fundamentally different from religious or moral claims? Why or why not?
8. Scientific research can be highly dangerous. Madame Marie Curie, for example, won the Nobel Prize but died from her work with radioactive elements. How does doing science, in this and several other respects, resemble service in the military? How does it differ from such work?

CHAPTER 3

The Spectrum of the Sciences

Science is not one great monolith but a collection of related yet distinct undertakings. Different branches of science have highly different methods, ranging from reproducible experiments to mathematical models to empirical investigations of events buried deep in the past. Because of its emphasis on historical questions, geology is unique among the sciences. The special tone of geologic research can be best appreciated by considering a specific example, followed by some remarks about the spectrum of science. We will begin with the quaint but important history of the interpretation of certain peculiar rocks.

The Example of Figured Stones

NOTHING IS SIMPLE

In the seventeenth century, as Oliver Cromwell's armies dethroned an English monarch and French kings bloodily repressed their peasants, scientists in these same two countries were fighting an entirely intellectual battle. Seventeenth-century naturalists[1] were fiercely debating the curious patterns

[1]In those days, the subfields of science were not clearly marked, so geologists were only one type of naturalist. Moreover, many gentlemen dabbled in what we now call physics, chemistry, geology, and biology. (Gentlewomen and lower-class people of both sexes were generally not permitted such recreation.)

found in some, but not most, rocks. These patterns, embedded in solid stone, were mostly small and obscure markings, but in a few instances they were more pronounced and clearly resembled the shells or bones of animals. In some rocks, the markings were detailed and clear yet completely different from those of any known organism. Some of the rocks even encased objects highly similar to the shellfish of the oceans; these were found in rocks near the shore and—much more remarkably—also in rocks far from the sea. The naturalists of the day called these remarkable discoveries *figured stones*, and they fiercely debated their nature and origin (Figures 3-1, 3-2, 3-3).

In retrospect, it is easy to see the clues to the organic nature of figured stones, but one must remember that the objects varied enormously in detail and kind. The naturalist's imagination also had something to do with the perception of some objects. At least, that is how historians explain the fact that the complete Latin and Greek alphabets were believed to be detectable in the markings of some of the stones. Other figured stones studied in the seventeenth century, however, clearly contained the complex details of marine and terrestrial life.

Some naturalists argued that the figured stones sprang from what we would call the inorganic world. They were not, these men insisted, related to organisms of the present or the past. In the inorganic realm, a force called **latent plastic virtue** had been postulated since Aristotle.[2] This power could account for the shapes found in all stones. The idea of a latent, creative power in inorganic substances was linked with the spontaneous generation of life, a notion accepted by virtually everyone from Aristotle through the nineteenth century. New life, such as frogs and worms, that emerged from the cold mud of lake bottoms each spring was created, it was thought, by spontaneous forces of the inorganic world. Likewise, those figured stones that closely resembled animals were accepted by many naturalists as examples of this formative force: "protoanimals," one might say, displaced into solid rock. The small and obscure figured stones were examples of the most primitive forms created by latent plastic virtue. Even more exotic shapes sprang from the same source, a tribute to the complex creative forces of the inorganic realm.

Another, opposing, group of naturalists believed that the figured stones were directly related to true organic life. Somehow, they contended, both animals and plants could die and be preserved in sediment that then turned them to stone. This, of course, is the view accepted today and taught to schoolchildren and visitors to natural history museums.

[2]This term might be rendered in modern English as "latent formative power." No less an authority than Aristotle had given the world the concept of a "plastic" or formative force in nature.

(A)

(B)

FIGURE 3-1: Which figured stone is of organic (biological) origin, and which is not? Geologists today think that all the objects in (A) are biological. Photo (B) compares a modern sand dollar (marine animal) with three pyrite "sand dollars" (of inorganic origin). (Photo: E. K. Peters.)

(A)

(B)

FIGURE 3-2: Which figured stone has a biological origin? It's not easy to tell. Geologists now believe that (A) is a stromatolite (an ancient marine algal mat) and that (B) is an inorganic iron ore (albeit with an organic-appearing texture). (Photo: E. K. Peters.)

FIGURE 3-3: This figured stone is known today as petrified wood. The surface shown here is the bark of a mature tree, but the sample is from a desert in the American Southwest that has no trees living on it. This is one example of the type of discrepant evidence that early geologists had to sort out. (Photo: E. K. Peters.)

SOME DIFFICULT QUESTIONS

Many seventeenth-century naturalists, however, had problems with an organic origin for figured stones.

1. How and why did relatively soft shells and bones become hard as rock?
2. Why were some figured stones, similar to modern marine organisms, embedded in rocks high in the mountains of Switzerland? Had the shellfish crawled across France, up the Alps, and somehow inserted themselves into the rocks?
3. How could one explain the figured stones that resembled the Latin and Greek alphabets? Why should they be dismissed as unimportant compared with the stones resembling animals?
4. Last, what about those figured stones that did not resemble any known creatures? Some of these peculiar stones had bizarre shapes and huge dimensions. How could they possibly be of organic origin?

> Neither the inorganic nor the organic hypotheses about figured stones that were propounded by seventeenth-century naturalists led to experimental testing. How, for example, could one directly confirm or disprove the organic nature of what we now call fossils? Geologists and biologists could hardly heap life-forms in a pile and then wait for years and years to see whether any of them would be preserved as figured stones.

But those who held to an organic view were tenacious. They developed the idea that mineral-rich fluids, such as those creating stalagmites and stalactites in caves, could replace organic material, bit by bit, with the substance of stones. They had never seen this process, nor could they validate it experimentally, but they nevertheless risked their scientific reputations on the idea. The marine fossils in the Alps, they contended, had been alive during a time when the oceans covered almost all of the earth. Many naturalists in the seventeenth century linked this idea with the biblical account of Noah and the flood. In any event, most naturalists began to believe that the sea level must have been much higher at some time in the past and then later dropped to its present levels. This conclusion would account for the otherwise strange topographic distribution of marine fossils. But once again, no direct observation nor any experiment could confirm the hypothesis.

THE CONCEPT OF EXTINCTION

Because some highly detailed figured stones resembled no known organisms, the French scientist Georges Cuvier (1769–1838) proposed the idea of the extinction of species. He thought that there had been quiet periods in geologic history when different animals and plants flourished and that these periods ended in mass extinctions triggered by violent geologic changes. This general view fit with the spirit of the Genesis account of creation if God rejected the earlier floras and faunas as not being "good." To put it crudely, God needed to make several tries to get what he wanted, discarding his earlier efforts at making plants and animals into the fossil record, perhaps using the agent of cataclysmic change over the whole earth to wipe out early examples of his handiwork.

Eventually, through discussion, speculation, and the discovery of more examples, all the questions about figured stones were resolved to the satisfaction of scientists everywhere. The debate was long and hard fought because it hinged on several untestable claims significant to larger questions about the history of life and the deformation of the earth's crust, ideas to which we will return later in this book.

How does the example of the figured stones fit into our larger concern, the full spectrum of the sciences? To answer this question, we must consider how different sciences are internally structured.

The Range of Sciences

PHYSICAL SCIENCE

Contrary to the assumptions of many members of our society, science is neither a monolith of testable hypotheses nor the collective work of objective, robotlike professionals. Rather, science is a rich collection of ideas that spans the spectrum of inquiry. As one Swedish physicist explained,

> Science is not a well-ordered, hierarchical, theoretical system. Fundamental dogmas, postulates, statements, or even results cannot be presented as basic characteristics of a "world view." Various sciences, instead, *form a turmoil of different and often contradictory ideas about assumptions, methods, and results.* [italics added] Although tempting, it is dangerous to believe in a unified, monolithic, scientific view of the world.[3]

Where does geology fit into the spectrum of sciences? One way to approach this question is to note that scientists from all disciplines generally believe that historical sciences like geology are derivative, or secondary, sciences, which rest on the fundamental physical sciences of physics and chemistry. It is commonplace to state that geology encompasses the scientific study of the earth and that geologists use any branch of physical science that is convenient. Indeed, the subfields of geology can be matched to the purer branches of science, as follows:

Physics: geophysics and tectonics (earthquakes and the movements of continents)

Chemistry: geochemistry and mineralogy/petrology (rocks and the chemistry of earth materials)

[3]Bengt Gustafsson, *The New Faith–Science Debate* (Minneapolis: Fortress Press, 1989), p. 1.

Biology: evolutionary theory and paleontology (the history of life and patterns of extinction)

Meteorology: hydrology and paleoclimate studies (the movement of groundwater; precipitation and weather)

Astronomy: geochronology (the age of the earth)

Physics and chemistry are disciplines in which hypotheses can be tested experimentally. In subatomic physics, the tests are extremely demanding and expensive. Furthermore, physics appears to be leaving behind the idea of testable hypotheses, in favor of "superstring" theory. Still, physics and chemistry are considered to be experimental sciences because in principle, the ideas in these fields can be tested. Despite the sophisticated experiments carried out by some geologists and biologists, the disciplines of geology and paleontology are still largely perceived as observational or **descriptive sciences** because they rely on a more passive study of nature.

A SECOND APPROACH

Another way of thinking about the differences among various branches of empirical work notes that a science like geology deals with a large number of entities and has a correspondingly large descriptive vocabulary. Although these branches of science do use deductive reasoning, logical deduction does not take up most of the researcher's time and energy. In addition, mathematics is used only sparingly.

A science like physics, on the other hand, handles relatively few entities, and its descriptive vocabulary is small. Instead, physicists devote much of their time to logical, deductive reasoning. As would be expected, their work relies on mathematics at every turn.

HISTORICAL SCIENCE

Almost all geologic problems are historical. Here we must digress. The phrase *geologic problems* is a traditional one and does not reflect the changes overtaking the broader field of earth science. Geology has traditionally dealt with such questions as the growth of mountain chains and the mapping of faults. These days, however, people who call themselves geologists also investigate the growth of the ozone hole and the chemical controls on contaminated groundwater. Clearly, geologists have expanded their field of inquiry, and in so doing they are now less bound to historical, rock-record problems. But for the purposes of this book—and for most introductory geology courses—geology investigates the history of the earth, broadly defined to include both physical and biological changes throughout our planet's long life.

Historical scientific problems often occur on a vast scale and over enormous stretches of time. Accordingly, evidence to support scientific claims can be reconstructed by logical reasoning related to observed data, but no experiments can demonstrate what actually happened in the geologic past.

THE METHODOLOGY OF THE HISTORICAL SCIENCES

One might think of the methodology of the **historical sciences** as similar to what physicians do. That is, the task in both medicine and historical science is to diagnose (understand) a particular case using generalizations gathered from many other cases. The analogy can be pushed further: in both medicine and natural science, every problem always contains an important historical component. For example, a patient's personal history can be crucial to understanding symptoms like anemia or a recurrent fever. Similarly, the geologic history of a mountain chain is key to understanding particular rock formations or knowing where oil might be discovered. Of course, the practice of medicine is not a pure science, by any stretch of the imagination. If historical science were analogous to what generally happens in doctors' offices, it would be far from rigorous. (And some people think that even the medical analogy may be too kind!)

One might also compare what most geologists do with the work of a police detective. Just as detectives investigate people and things in the present and try to understand what occurred at the time of the crime, so geologists look at present-day rocks and fossils and try to understand events in the past. As one popular book puts it: "[G]eology, like criminal investigation, deals with events which have already taken place and works to reconstruct the order and cause of these events, using methods more speculative than testable."[4]

Perhaps more remarkably, it is not a criticism to say to a geologist who is explaining ideas about a set of data: "Well, yes, that's a good story you've got there." Indeed, we think that geologic history is a succession of such stories. There are, of course, rigorous and peer-reviewed constraints on the structure of the story a geologist can put forward, but the idea of **narrative** is woven throughout the work of historical science. Putting the two ideas together

[4]John E. Allen and Margorie Burns, with Sam C. Sargent, *Cataclysms on the Columbia* (Portland, OR: Timber Press, 1986), p. 12.

leads to this thought: if geologic work contains a large element of storytelling, the most obvious type of tale to construct would be the detective story. This image is occasionally taught to introductory students: "Field geologists function as detectives. The acts of nature usually have been completed long before they arrive on the scene. Their task becomes one of reconstructing events . . . start[ing] with the most recent event and work[ing] back in time. Sherlock Holmes called this method thinking backwards."[5]

Few scientific disciplines would allow their most important researchers to be compared with a fictional Victorian detective. It is one of geology's virtues that its practitioners occasionally are protected from the intellectual elitism of some other branches of scientific research.[6]

HISTORICAL SCIENCE IS NOT TIDY

By its nature, geology has to grapple with the distant past and many different types of evidence. These characteristics of geology are illustrated by the debate about figured stones. Some parts of science, notably traditional physics, converge toward elegantly simple fundamentals. In contrast, geology is faced with dispersive tasks and specialized exercises in narration.

> Despite the untidiness of geology, researchers have made enormous progress in understanding the earth's complex features, ranging from figured stones to mountains. Geology has advanced because several generations of geologists have done good work, even if that work does not fit comfortably the stereotype of neatly objective and testable scientific methodology. This geologic research has revolved around discussions—one might even say vociferous arguments—that have been marred by personal bias but that overall have allowed new and sweeping understandings of the earth. These arguments have moved forward by fits and starts, quite unpredictably, and have often been suspended for decades.

[5]John W. Harrington, *Dance of the Continents: Adventures with Rocks and Time* (Boston: Houghton Mifflin, 1983), p. 21.

[6]Interestingly, and employing quite a different tone from this book, the philosopher Robert Frodeman offers geological science for our consideration as the *queen* of different branches of intellectual inquiry. The many different methods and types of thought that geologists use, Frodeman argues, makes them uniquely qualified to address the problems of the twenty-first century. Some of those who took a "Rocks for Jocks" course might be surprised to learn about this perspective, but it is clearly set out by Frodeman in "Geological Reasoning: Geology as an Interpretive and Historical Science," *Geological Society of America Bulletin*, August 1995, pp. 960–968.

Philosophers and historians of science, like Thomas Kuhn and his intellectual descendants, generally concentrate their studies on physics. They have done a great deal of work emphasizing periods of "revolution" in which fundamental **paradigms** are disputed and a new view emerges, followed by periods of "quiescence" in which research consists of filling in the gaps of the existing paradigm. For example, once Newtonian mechanics (force = mass × acceleration and all the other topics of high school physics) was established, physicists did not question its assumptions. Rather, they worked within the Newtonian framework to flesh out other areas in mechanics like the conservation of angular momentum. In the vocabulary of Thomas Kuhn, the Newtonian "paradigm" reigned supreme for generations. But after that long, quiet period, tensions built up between the old paradigm and new thoughts about light and gravity, about individual components of the atom, and about the relationship of time and space. This revolutionary period led to Einstein's general theory of relativity and to the postulates of quantum mechanics. We now do our scientific work within these paradigms.

Not everyone, however, accepts Kuhn's model. More recently, another idea has been put forward, one using an analogy between the progress of science and Darwinian evolution. According to this view, scientific theories evolve through generations of researchers, owing to the creative forces inherent in the "withering" criticism to which such theories are exposed. Good theories are gradually modified through time, not in sudden revolutions, but in the mundane world of peer review and scientific conferences.

Geology and paleontology do not fit easily into either of these accounts. The history of these sciences is often argumentative, and it may involve withering criticism, but it is also wonderfully untidy and gloriously disparate. In fact, research in the historical sciences is almost as disorganized and contradictory as American politics, even though many of the big questions are never disputed. Often, major lines of inquiry are simply bypassed by geologists and paleontologists interested in another set of ideas.

It may be that historical science does not fit easily into the orthodox views of science in part because it is principally a narrative art. Geology is different from physics, chemistry, and even most of biology. Furthermore, geologists usually can work on interesting problems without embracing overarching theories; much of what geologists do is tied to specific localities and unique events in history. In other words, "the history of geology, like the history of the Earth, follows a comprehensible and explicable pattern but changes in different ways, at different times, and with different intensities, in a manner that hypotheses of linear progress or alternating stages can only caricature."[7]

[7]Mott T. Greene, *Geology in the 19th Century* (Ithaca, NY: Cornell University Press, 1982), p. 294.

SUMMARY AND CONCLUSIONS

Science encompasses many different activities. Geology and paleontology are historical sciences, concentrating on deciphering the earth's rock and fossil record. Because it investigates events buried deep in the past, traditional geology uses observation and inference rather than experiment.

Good observational work eventually established the modern view of fossils. But in the seventeenth century, there were sound reasons to believe that figured stones must have an inorganic origin. The problem, of course, was that questions about fossils rested on questions about the history of life, the rise and fall of the oceans, and the deformation of the earth's crust. Progress in all these areas was necessary to resolve the figured-stone debate.

QUESTIONS FOR DEBATE AND REVIEW

1. How would Aristotle have explained the origin of figured stones? Were his views scientific?
2. What branches of science are known for experimental work? What branches of science are known for mathematical modeling (which cannot always be experimentally verified)? What branches of science are best termed historical?
3. Consider again the problems that an organic origin for figured stones entailed. In the seventeenth century, was it reasonable for some good scientists to be on each side of the figured stones debate? Why or why not?
4. Police detectives and the specialists they hire do a lot of empirical work (analyzing bloodstains, fingerprints, bomb fragments, and the like). Are detectives really scientists? If not, and given the analogy we drew between geologists and detectives, are geologists really scientists? Explain your answers.

CHAPTER 4

Neptune Versus the Fires of Hell: The Dynamic Cycle of Minerals and Rocks

If the origin of fossils was a difficult debate for generations of naturalists, it will come as no surprise that the origin of some rocks was also problematic.

A few rocks are clearly sedimentary; that is, they are formed at the bottom of lakes, rivers, or oceans. Sandstone, for example, looks just like beach sand that has somehow been cemented together into solid rock. Similarly, shale and mudstones look like compacted clays and muds. Early geologists throughout Europe agreed on a sedimentary origin for these rocks, a view that encouraged the growth of **Neptunism**, or the assumption that virtually all rocks formed in the ocean. (Neptune was the god of the sea in Roman mythology.) To make Neptunism square better with the Christian tradition, early geologists often assumed that most sedimentary rocks had formed during the worldwide flood of Noah's day, on what is now land.

But were all rocks formed by sedimentary processes? Like the mystery of figured stones, this was a difficult matter to investigate experimentally. But good observational work and sound reasoning eventually led to the modern view of the origin of rocks, a view heralded in traditional geology texts as the **rock cycle**.

Understanding the early debate about one rock in particular will help us see how geologic research is done. We will then expand our terminology with the modern view of the origin of many different rocks.

The German View of Basalts

Rocks that looked sedimentary in origin—such as sandstone, shale, and fossil-rich limestone—were not controversial. In Germany, early geologists studied large, horizontal beds of these rocks and mapped them across extensive areas of their country. Occasionally, however, interbedded in the sedimentary strata, a different type of rock was found, called basalt.

Most samples of basalt are hard, dark brown-black, and rather featureless (Figure 4-1). At first glance, the basalts looked like shale or mudstone, usually with no individual mineral grains evident. But basalt is much harder than shale, and German geologists quickly realized it was a distinct type of rock.

Since the basalt beds in Germany were horizontal, sandwiched between clearly sedimentary rocks, the geologists there considered them also to be sedimentary in origin (Figure 4-2). Basalt, they reasoned, was a chemical precipitate from an ancient ocean or perhaps from the floodwaters of Noah's time.

Then, as now, chemical precipitates were known to form from salty solutions. In deserts, for example, shallow bodies of water evaporate in the sun and leave behind mineral precipitates. In a similar manner, the German geologists

FIGURE 4-1: Black, featureless basalt rock. Such samples can be difficult for a novice to distinguish from mudstone and other dark sedimentary rocks. (Photo: E. K. Peters.)

FIGURE 4-2: This river has exposed horizontal beds of basalt similar to those seen in Germany. The horizontal rock beds suggest a sedimentary origin. (Photo: E. K. Peters.)

reasoned, basalt had formed slowly at the bottom of an ancient ocean. They noted that some basalt outcrops displayed vertical columns in their thick layers (Figure 4-3). In cross section, these basalt columns were roughly hexagonal. To the Germans the shapes were reminiscent of the polygonal forms that appear in mud after it dries. Again, a sedimentary view was confirmed, and Neptunism seemed like the best approach to determining the origin of most rocks.

There were a few disturbing features about basalt, however, that did not fit well with this theory. Occasionally, basalt rock in Germany was discovered in thin vertical veins rather than in thick horizontal beds (Figure 4-4). Such shapes were more closely associated with the veins found in the rocks near volcanoes in Italy. It seemed that basalt could, at times, behave more like liquid lava than a chemical precipitate, squirting up into cracks in surrounding rocks.

But the German geologists noted that these veins of basalt were rare, and they decided that the few places where this volcaniclike basalt occurred indicated that deeply buried basalt rock could be melted by localized pockets of heat in the earth. The source of these heat pockets, they speculated, was coal beds that were on fire. This idea was grounded firmly in experience, as coal beds found in sedimentary rocks were being mined in Germany by this time.

FIGURE 4-3: A thick, horizontal basalt bed is cut by vertical columns or cracks (see the power sign for scale). (Photo: E. K. Peters.)

All too often a spark ignited the coal gases in the mines, and the fires underground sometimes burned for months and years.

Such subterranean fires, the geologists reasoned, could melt a little sedimentary basalt. The liquid then moved upward along fractures to form the basaltic veins. These small areas of liquid basalt behaved like a volcanic lava, although such anomalies did not mean that basalt was not fundamentally a sedimentary rock precipitated from an ancient ocean.

(A)

(B)

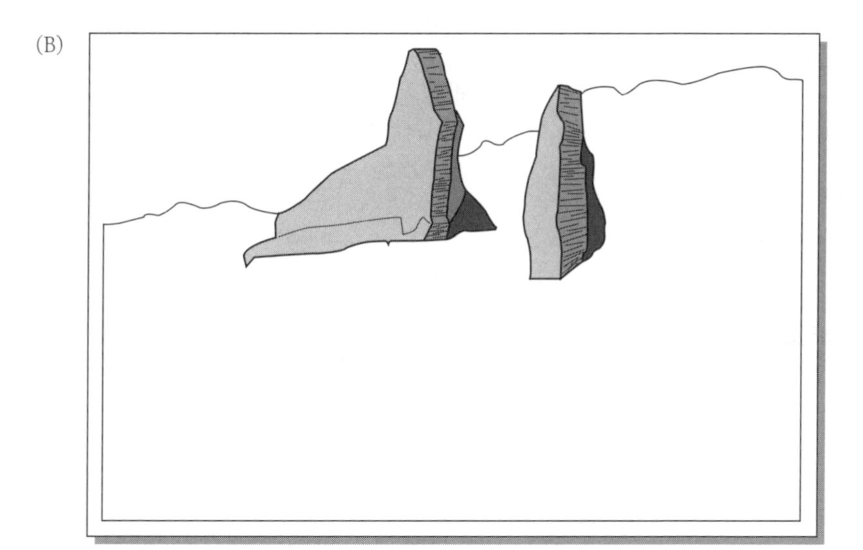

FIGURE 4-4: These two vertical veins of basalt show the type of formation that impressed English geologists and did not fit superbly with the German view.

The English View of Basalts

Geologists working in England and Scotland took an entirely different view of basalts. They mapped basalt outcrops far removed from coal beds and far from any undisputed sedimentary rocks. Basalt veins in England and Scotland cut across granite, a rock that everyone agreed was not formed in a sedimentary environment. Not surprisingly, English-speaking geologists considered basalt to be **volcanic**, pointing to the undisputed volcanoes and volcanic rocks in Italy as analogues of what they discovered in England. The rocks in Italy were not exactly basalts, but they were closer to basalts than they were to any clearly sedimentary rock.

With help from their French counterparts, the English geologists attacked the views of their German colleagues. They were not impressed by the thick beds of basalt found in sedimentary sequences in Germany, in part simply because they had little acquaintance with them. Such basalts, the English and French reasoned, had squeezed as a liquid in between preexisting sedimentary rocks. Critics in Germany asked what force could be responsible for such an intrusion, an event that would have had to raise all the sedimentary beds above the basalt. This perfectly good question was, indeed, difficult to answer.

On firmer ground, the English pointed out that basalt layers never contained any fossils. The vertical columns in the basalt were not formed by drying, as German geologists had reasoned, but by the cooling of molten lava. As the molten basaltic rock cooled and solidified, it contracted, as liquids generally do when they become solids, and thus the vertical columns were formed. Contraction due to cooling, rather than drying, was the key, according to the English.

Volcanism Wins

As more and more field evidence about basalts was gathered, it became increasingly clear that a volcanic origin for the rock best fit the facts. There were, indeed, no fossils to be found in basalt. Moreover, some chemically identical rock was found full of holes (Figure 4-5). These holes are called *vesicles*, and rocks full of vesicles were found in and around volcanoes in Italy. Finally, in Iceland, samples of basalt could be found in the mouths of active volcanoes.

English and Scottish geologists, triumphant on the subject of basalts, proposed an igneous (or molten) origin for granite as well. The evidence regarding the age and origin of granite is more complex than we can describe here, but suffice it to say that in time, molten rock was accepted as the origin for granite just as it had been for basalt. This view, which we shall simply call volcanism, required that the earth's interior contain significant amounts of molten material. The volcanists did not have an explanation for why such melts should be present. They proposed a fiery layer beneath the earth's surface, but they could not prove, or explain, their proposition. Still, the volcanic nature of basalt had become clear to all concerned, and since there was

FIGURE 4-5: Basalt sample with holes (vesicles). English geologists thought that such samples looked volcanic, and they argued that such holes were never found in sedimentary rocks. (Photo: E. K. Peters.)

so much basalt, it seemed that there must indeed exist subterranean masses of molten material. That this assumption fit with the Christian notion of a hell located somewhere "downward" no doubt helped some geologists of the day accept the igneous view of basalts and granites.

None of this means that the early Neptunist geologists working in Germany did poor work. In fact, many features of the basalts they studied made good sense in a sedimentary framework (Figures 4-6 and 4-7). Later, it became fashionable to dismiss the Neptunists as biblical literalists, linking them with Noah and the flood as a supposed explanation of everything around us. But the German geologists were much more sophisticated than that: They were Neptunists because the evidence favored a sedimentary view of the basalt layers they found across their country. The best geologists among them thought that the rocks had formed not in a catastrophic flood but as a slow precipitate from an ancient ocean.

As it happened, other samples and other features of basalt were found in other parts of Europe, and in time, the volcanic nature of basalt was accepted by all geologists. The debate was a long one, but it eventually was resolved by

FIGURE 4-6: This outcrop looks like shale, even to experienced geologists today, but is really made of basalt which has weathered in an unusual way, producing the horizontal features associated with sedimentary rock. (Photo: E. K. Peters.)

the weight of the evidence from many different countries. Today, geologists can confirm the igneous origin of basalt by watching active volcanoes in Hawaii. Scientists can also create basalt in the laboratory by melting mixtures of silicate material at high temperatures and allowing the melt to cool rapidly.

One Man's Religious Assumptions Prove Surprisingly Helpful

James Hutton (1726–1797), sometimes called the father of modern geology, was a Scottish naturalist who excelled at fieldwork and logical reasoning. A careful examination of sedimentary rocks convinced him they had been formed by sediment derived from the erosion of preexisting rocks. Studying the streams and beaches near him, he noted that erosion and sedimentation were exceedingly slow processes. This meant, reasoned Hutton, that the earth was very old. Only a great span of geologic time could enable the formation of sedimentary rocks.

Hutton's sound argument for the earth's antiquity was a step forward from the ideas about the age of the earth then current in England. Most English

FIGURE 4-7: Samples A and B are organic-rich shale. Sample C is basalt that has naturally weathered into shale-like tabular forms.

speakers accepted the notion that the earth was as old as a literal reading of the Hebrew Scriptures allowed. Thus when Hutton's work became understood—some years after his death—English geologists rapidly agreed that the earth was enormously old.

Hutton made another major point, one that may seem less reasonable to us today but that was central to his thinking. Hutton was a theist who believed in a creator God, and he believed that the complexity and order in nature reflected the goodness of an all-powerful creator. This theological idea, called the **argument from design**, dominated a great deal of English intellectual life until the beginning of the twentieth century. The British are known for their devotion to this argument, one branch of which is known as *natural theology*. It was because Hutton took seriously the argument from design that he assumed that the earth, unlike an unwound clock, would not stop working.

Since erosion always wore away rocks and land and shifted their sediments to the sea, Hutton saw that the world could not remain as it was unless a restorative process created more rocks on land. This restorative force that Hutton sought could be found, he reasoned, in igneous rocks. Earlier geologists had thought that granite was primordial (unchanged from the beginning of time), whereas Hutton argued that granite and other igneous rocks were produced in the earth in processes continuing throughout earth history.

When these igneous rocks were lifted up high onto the surface of the land, they supplied the material that erosion could wear away to create sedimentary rocks. There was, Hutton contended, a balanced cycle at work. Igneous rocks were created and uplifted at the same rate at which erosion wore them away and created sedimentary rocks.

From our perspective, Hutton's theological views led by chance to a fruitful scientific idea, one that stands at the center of modern geology. From Hutton's perspective, of course, the argument from design was helpful because the world was, in fact, created by a benevolent God.

Today there are many lines of evidence, including experimental work in laboratories, that persuade geologists to believe the earth's rocks have three fundamental origins linked together in a continuing and dynamic cycle.

The Rock Cycle Today

Rocks formed from molten material are called igneous rocks. Those igneous rocks that cool at or near the surface crystallize fairly rapidly and are labeled volcanic. Basalt is one such rock. Those igneous materials deep beneath the surface cool slowly, allowing large crystals to grow. Such rocks are labeled plutonic, and granite is an example of plutonic rock. Granite contains crystals visible to the unaided eye, whereas in basalt the crystals are generally visible only through a microscope.

Virtually all molten masses in the earth, and therefore the igneous rocks derived from them, are rich in silica (one silicon atom bound to oxygen atoms). Silica-rich rocks are abundant simply because silicon and oxygen are the most common elements in the earth. Those igneous rocks richest in silica are generally light colored and are termed **felsic**. Those containing a bit more iron and magnesium are generally dark colored and are called **mafic**. Both mafic and felsic igneous rocks can be found in the volcanic or the plutonic realms.

When igneous rocks are exposed to rain, snow, wind, and frost at the earth's surface, they undergo a process called *weathering*. This breaks down the rocks, both physically and chemically. Rocks are composed of minerals, such as quartz (pure silica), garnet, and mica, as well as many other, less familiar minerals. Two other minerals abundant in the earth's crust are the feldspars (a silica-rich mineral found with quartz in granite) and calcite.

Calcite is the only mineral mentioned here that does not contain silica. It is instead a carbonate mineral, made of calcium carbonate. Seashells are made of calcite, and calcite can also precipitate in the ocean by means of inorganic chemical reactions.

To return to the concept of weathering, small particles of igneous rocks, individual mineral grains, and dissolved chemical constituents are washed by streams into rivers and oceans, where they form thick layers of sediment. In the marine environment, familiar sedimentary rocks such as shale and sandstone (Figure 4-8) are formed from this sediment. Limestones are generally created by biological activity that traps calcium carbonate in the shells of sea animals. When those shells accumulate in beds, fossil-rich limestone is formed.

The third origin for rocks is deep in the earth, especially in zones of mountain building. We call this realm metamorphic. Experiencing great heat and pressure, igneous and sedimentary rocks are deformed. Because of the enormous pressures in the earth, these rocks do not truly melt but become a bit like Silly Putty or warm wax, deforming slowly through staggeringly long periods of time. The textures of metamorphic rocks often are dominated by bands or swirls (Figure 4-9). Marble is a familiar metamorphic rock derived

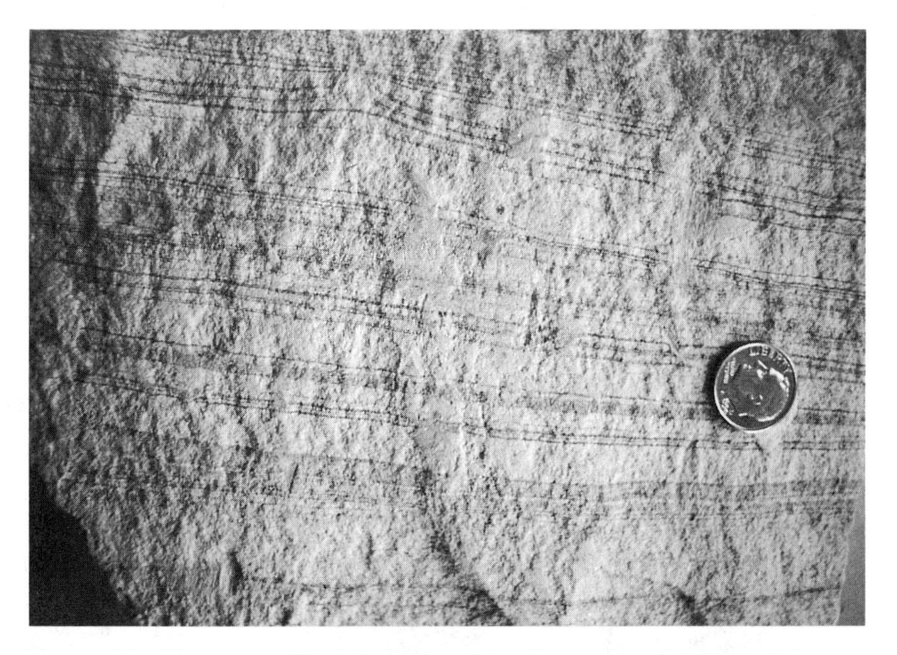

FIGURE 4-8: The texture of finely bedded sandstone betrays its sedimentary origin. (Photo: E. K. Peters.)

FIGURE 4-9: These building stones are made of gneiss (pronounced "nice"). The metamorphic origin of gneiss is revealed in the rock's texture. (Photo: E. K. Peters.)

from limestone, whose texture is formed by the plastic deformation the rock has experienced in the earth. When a rock like shale is metamorphosed, it becomes slate, and if the metamorphism is intense enough, rocks called schist and gneiss will form. Gneiss can also be created from the intense metamorphism of granite. The texture of gneiss, like marble, betrays the deformation produced by the metamorphic process. Finally, when sandstone is metamorphosed, the quartz sand grains are fused together by silica, forming the dense but light-colored rock called quartzite. Quartzite is the rock most resistant to weathering and therefore should be the choice for tombstones of those who want their names remembered for millennia.

> To sum up the rock cycle: The weathering products of igneous rocks eventually become sedimentary rocks. Both types of rocks can become metamorphosed if they experience great heat and pressure in the earth. Metamorphic rocks may melt, passing into the igneous realm. Or metamorphic rocks may be uplifted and eroded into sediments. The basic elements of this cycle are displayed in Figure 4-10.

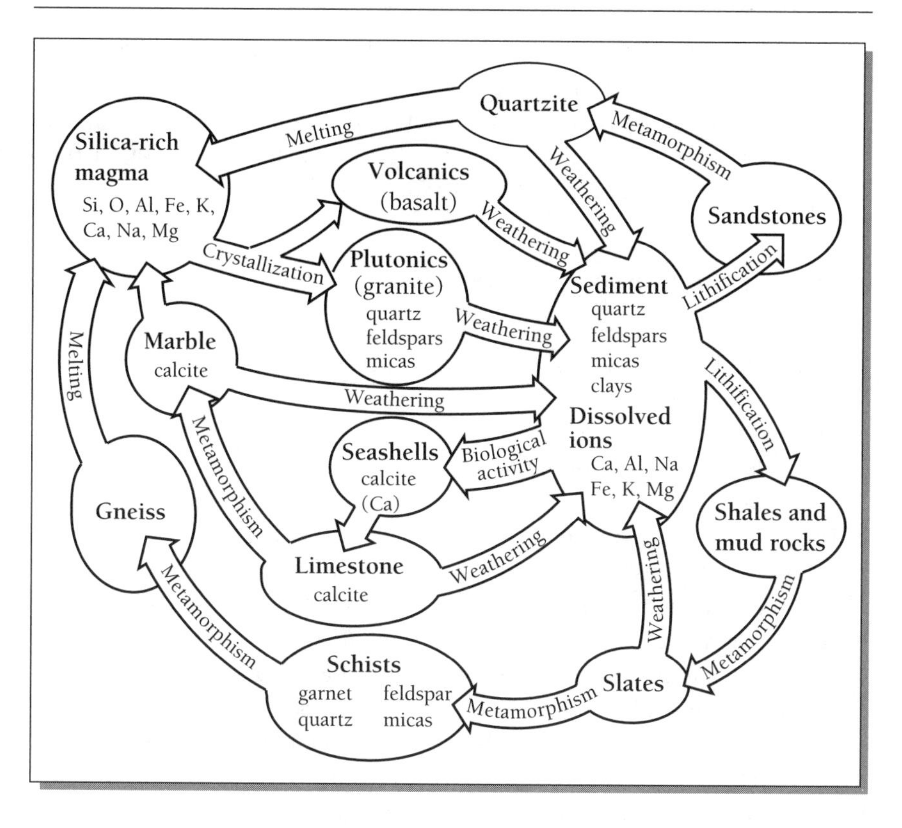

FIGURE 4-10: The rock cycle. (Figure developed by L. Davis.)

SUMMARY AND CONCLUSIONS

The origin of the rock called basalt was a subject of heated debate among early geologists. In Germany, there was good evidence that basalt was a sedimentary rock, which led to the view of rock formation called Neptunism. Neptunists were not, on the whole, biblical literalists enthralled with the story of Noah and the flood. However, their opponents liked to link them to such thinking in an effort to discredit good field evidence in Germany. In France, Italy, England, and Scotland, different types of evidence regarding basalt were revealed. In time, geologists across Europe accepted the idea that basalt was volcanic.

James Hutton conceived the basic idea of a rock cycle, linking the genesis of one set of rocks with others. His assumptions about the nature of a creator God led him to a highly productive idea about igneous rocks being continuously created. Such rocks, he argued, were lifted to the earth's surface, especially in mountainous areas. There they were exposed to agents of weathering and slowly turned into sediment.

The rock cycle is our modern expression of the core of Hutton's idea, and its vocabulary is studied in all introductory geology classes as the foundation for our understanding of earth processes.

QUESTIONS FOR DEBATE AND REVIEW

1. You have no doubt heard about the volcanoes in the state of Hawaii. What kind of rock do they erupt? In the tropical climate of the Hawaiian Islands, do you expect that weathering processes are, geologically speaking, rapid or slow?
2. What does the nature of the rock cycle say, if anything, about the age of the earth? Could a very young earth (created a century ago, say) have sedimentary and metamorphic rocks? Explain your answers.
3. Name the rocks used as building stones near where you live. Hints: Look for limestone (identified most easily when it contains fossils) and marble (usually light colored and with a marbleized texture). You many find granite or gneiss in the building stones of banks. On the East Coast, a few old sidewalks are made of slate. The foundation stones of old buildings are often sandstones. Be sure you can place each rock in its proper place in the rock cycle.
4. Does James Hutton's reliance on the argument from design fit well with our view of the scientific method? Explain. Was his approach useful in this particular case?

Chapter 5

Uniformitarianism and Gradualism: Geology's Shifting Foundations

It is often said that **uniformitarianism** is the founding principle of geology, and textbooks for introductory courses often devote several pages of the first chapter to discussing uniformitarianism and related concepts. The idea behind the term is usually credited to James Hutton (1726–1797), and many textbooks explain that a later geologist, Charles Lyell (1797–1875), developed and applied the concept.

A simple example illustrates the idea behind uniformitarianism. Today, high in the Rocky Mountains, geologists can study *glaciers* (moving masses of ice) as they slowly flow down alpine valleys. They can study *moraines* (piles of rock debris pushed along by the glacier) and *striations* (deep grooves cut into bedrock by stones caught in the bottom of the ice as it flows). Although these features are currently being formed, those created in the past allow geologists to interpret their similar features. Geologists are confident about their conclusions even in regard to moraines and striations where no glaciers are present today, such as in Scotland or South Africa. A phrase that geology textbooks often use to sum up the concept of uniformitarianism, and therefore the spirit of geologic inquiry, is "The present is the past."

The term *uniformitarianism* has several uses, and despite the buoyant tone of introductory textbooks, most geologists give the concept contradictory meanings. We need to look at scientific history to better understand our current use and misuse of geology's foundational principle.

The History of the Idea of Uniformitarianism

CATASTROPHISM

When geology was in its infancy, many people assumed that the earth had been shaped by violent forces that operated only intermittently. The eruption of volcanoes and the instantaneous shock of earthquakes fit with this view, which often ascribed responsibility for these violent events to divine influences. Those people who used this framework in their study of the earth are termed **catastrophists**. In parts of the biblical account of earth history, God's actions are clearly catastrophic. The great flood of Noah's time is a prime example of sudden and violent change across the face of the earth. The earliest English naturalists assumed that earth history was both catastrophic and short. The Church of England, citing biblical authority and ecclesiastical calculations, advanced the year 4004 B.C.E. for the earth's creation.

Catastrophism was also an assumption of the later, undisputedly brilliant, French scientist Georges Cuvier. As we saw earlier, Cuvier argued that fossils represented past living creatures. He believed that floras and faunas quite different from our own had flourished in the past, only to be wiped out by catastrophic changes that led to mass extinctions.

HUTTON AND LYELL

Hutton and Lyell, the two heroes of traditional geology textbooks, rejected both catastrophism and the young earth demanded by the common thinking of their day. They developed a different set of assumptions for earth processes, contending that geologists should study currently active and gradual processes. They pointed to such ongoing events as the erosion of soil during spring rains and the accumulation of sediment in lakes. Next, Hutton and Lyell believed, geologists should apply, as directly as possible, lessons from current processes to the geologic past.

As it happens, most of the change we see around us is relatively gradual. For instance, over and over, we see small-scale erosion on hillsides in the spring but only rarely do we experience an earthquake. When geologists focus their thinking on common, gradual processes, they are subscribing to what is termed **gradualism**. As the name suggests, gradualism is the belief that the changes recorded in earth history reflect gradual processes. Hutton and Lyell argued that by extrapolating from the gradual erosion they saw, they could imagine how large valleys and deep canyons were carved out through a long span of time (Figure 5-1). The notion that the important processes shaping the earth are gradual demanded that geologists expand their reckoning of its age.

Notice that a geologist could embrace uniformitarianism but not believe that all changes on the planet were gradual. Such a person might emphasize volcanic eruptions, earthquakes, and other cataclysmic processes. However, the two ideas of uniformitarianism and gradualism easily complement each other. The early geologists who were gradualists were also fond of the concept of uniformitarianism and believed in an enormous age for the earth. They were, generally speaking, opposed by the catastrophists, who continued to accept a shorter history punctuated by a few spectacular events not necessarily represented by current processes.

It is tempting to dismiss catastrophists as biblical literalists enthralled by the flood and the dramatic parting of the Red Sea, men who did not understand the modern, empirical spirit of science. But the important catastrophists, like Cuvier, were far from devout and far from antiempirical. Nevertheless, in the eighteenth and nineteenth centuries, geology advanced as a discipline principally when geologists assumed a gradualist framework of thought.

FIGURE 5-1: This canyon in the Pacific Northwest is most easily explained as a feature formed over a long span of time by gradual erosion. (Photo: E. K. Peters.)

GEOLOGY AND THE RELIGIOUS AUTHORITIES

The difficulty between geologists and the church authorities rested, as it so often does, on different background assumptions and different resultant methodologies. Bishop Ussher (1581–1656), speaking for the Church of England, had assigned the earth a birthdate in the autumn of 4004 B.C.E. The date was based mainly on an interpretation of Holy Scripture and was accepted by generations of educated English men and women. English geologists, in particular, had to combat this dogmatic view to make any progress interpreting the planet's historical record. The great *depth*, as we call it, of geologic time was apparent in qualitative terms to Hutton and Lyell, and it clearly contradicted the church's teaching about the age of the earth.

The rocks that Bishop Ussher could have inspected—if he had cared to look—included sedimentary strata. In some parts of England these rocks are stacked horizontally, one atop the other, for thousands of vertical feet. Hutton thought of the sand and mud he had seen accumulating at the bottom of great rivers and at the ocean shore, and he was led to suppose that thousands of feet of such sediments took more than a few years to be deposited. Bishop Ussher and his supporters insisted that all such rocks had formed during Noah's flood.

It is easy to forget the early geologists' struggles with the general intellectual culture around them. Like Charles Darwin many years later, the founders of geology came up against the assumptions of Christian thought as it then existed. The division was sharp and deep: Bishop Ussher's followers were not interested in the empirical evidence of sedimentary rocks and erosion because they looked to scripture (revealed truth) to explain the natural world.

> Despite their struggles with religious leaders and secular catastrophists, the gradualists forged ahead, and gradualism eventually became the accepted assumption behind most of the work pursued in the geologic community. Because gradualism was a constructive approach to many problems, it came to be embraced by the discipline as a whole. Nevertheless, so much intellectual and emotional energy had been spent fighting biblical literalism that geologists developed an automatic hostility to any catastrophic hypothesis, however secular.

GRADUALISM GAINS GROUND

Even in the middle decades of this century, the abhorrence of catastrophic change in geologic history was great enough that introductory textbooks did not even mention the word. And because the idea of catastrophism was completely ignored, gradualism also was not discussed! That is, the authors of introductory texts presumed a gradualistic view of earth history, and innocent undergraduates did not know enough to question what they were being taught. Today students of introductory geology know that gradualism is still invoked by geologists. For example, the Colorado River, geologists say, carved out the Grand Canyon by removing tiny bits of soil and rock each and every day for a million years. But there also is room in a uniformitarian view for some episodic and dramatic processes, such as volcanic eruptions. Geologists can see, albeit rarely, catastrophic processes today, and so they can assume that in the past these occurred in a similar manner, even if with a different intensity.

> Most geologists believe that although the rates and intensities of geologic processes may have varied in the past, the processes themselves were the same as those seen in action today. This idea is often taken to be the modern expression of uniformitarianism.

The Foundation of Scientific Work

METHODOLOGY: WHAT WE ASSUME VERSUS WHAT WE DISCOVER

Many statements about uniformitarianism contain serious shortcomings: Geologists have no empirical reason to think that the instant of geologic time in which they live is representative of the earth's past. Furthermore, they must admit that some processes operating in the past might not even be active now. Indeed, virtually all geologists accept that this is the case for at least one important event in earth history: the origin of life on this planet.

Geologists think that no life-forms existed in the earth's earliest history when temperatures were very high and the atmosphere was composed of carbon dioxide, ammonia, and hydrogen sulfide. Sometime during this period, roughly 4 billion years ago, scientists believe that life arose from inorganic molecules. The spontaneous generation of life is a process we do not see occurring in the world around us today. Thus, the most fundamental processes in the history of the earth, the events leading to life, do not agree with the principle of uniformitarianism as it is often expressed.

To remedy this difficulty, geologists can change their definition of uniformitarianism to mean the assumption that the laws of nature were followed in the past in the same way as they are today. Here scientists seem to stand on firmer ground. Such a definition has recently been proposed in a standard geology text.[1] The difficulty with this approach, however, is that uniformitarianism begins to lose content. Saying that the laws of nature operated in the geologic past in the same way as they do today is an assumption about earth history, not a description of it.

When scientists make a claim that can be disproved by experiment or observation, that claim is termed *substantive*.[2] "The earth revolves around the sun" is a substantive claim because one can imagine data that would conflict with this heliocentric view. Indeed, much of the work in the early history of science addressed this question. It was eventually resolved by numerous observations taken as a falsification of the previously accepted notion that the sun revolved around the earth.

In contrast to the idea of substantive claims, many of scientists' assertions are **methodological.** Such assertions state the background assumptions on which science operates. One is termed the **Principle of Parsimony,** which states that if a person can imagine two explanations for the same phenomenon and both logically accommodate all the available data, the simpler of the two explanations is preferable. One form of this idea originated with the medieval theologian William of Ockham (1285–1349), who is often credited by practicing scientists with the idea labeled **Ockham's razor.**[3] Notice that there is no way to falsify a methodological claim such

[1]Sheldon Judson and Steven M. Richardson, *Earth: An Introduction to Geologic Change* (Englewood Cliffs, NJ: Prentice-Hall, 1995), p. 535.

[2]This is a simplification. Much of geological theorizing cannot be falsified because geology is a historical and descriptive science. Nevertheless, we still can say that geologic theories are substantive. Not all our ideas may be falsifiable, but we test them by how well they fit with other (sometimes falsifiable) theories and with overarching frameworks based on observation and experiment.

[3]William of Ockham had no empirical basis for promoting the Principle of Parsimony. His position was a theological, not a scientific one. He believed the natural world must reflect the mind of God (who created the world). For Ockham, like many intellectuals dating back to the Greeks, simplicity and elegance were the highest of values. God, as a perfect being, would reflect only such values, and therefore the natural world (including the human mind) must be fundamentally simple.

as the Principle of Parsimony, because methodology is not substantive. It is, rather, a system of rules that scientists choose to follow about how science should proceed.

One can imagine some alternatives to Ockham's razor (the Principle of Parsimony). Perhaps when confronted with two adequate hypotheses, scientists should prefer the more *complex* one, on the principle that nothing in the world looks simple! Then again, perhaps scientists should prefer the explanation or hypothesis that *first* occurs to them. Such a principle might be that intuition is an unconscious and richly creative path toward new knowledge. Still again, scientists might prefer the explanation or hypothesis that weighs most heavily on them after deep, silent *contemplation*. In this case, the principle would be that by means of contemplative reflection we will choose the best and truest explanation. The point is that Ockham's razor, although invoked by almost all scientists as if it were substantive, is in fact simply a choice about how scientists will proceed with their work, and they could just as easily make other choices. In other words, at least in principle, the day may come when scientists collectively decide to pursue their work using a different methodology.

THE HEART OF THE MATTER

A fundamental question that science must grapple with is whether the laws, or constants of nature, have been the same through time.[4] Scientists have chosen to assume that the regularities we observe do not change through time. The values of certain constants in physics, for example, are taken to be the same today as they were in 1776 or as they were 100 million years ago. One could imagine a different scenario, and it would be impossible to prove one way or another, for at least some changes in the universe's physical constants would not be recorded in data in a way we could decipher. But despite our ability to conjure up such nightmares, as scientists we select the simplest or most evident view, that physical constants have remained the same through time.

[4]Scientists often talk about the "laws" of science, such as the law of gravity or the three laws of thermodynamics. The term is curious to the modern reader, for clearly the world need not follow "laws" laid down by a court or legislature. The idea of scientific law rests on the explicitly religious notion that God's creative intentions are shown by the regularities of the natural world. Early scientists were theists (they believed in a creator God), and they believed that the natural world followed God's chosen "laws" of creation. Today, of course, scientists need not link the regularities they observe to divine will, but the term law is still in use. For our purposes, it is enough to say that scientific "laws" are ideas that scientists would bet their lives on, whereas they hold to "theories" and "hypotheses" with much less certainty.

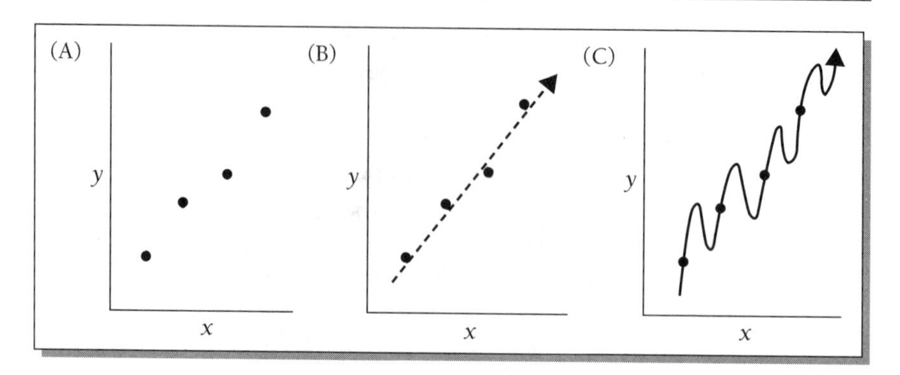

FIGURE 5-2: (A) A diagram of data gathered on two types of measurements. (B, C) Two interpretations of the data shown in (A).

Here is a generalized example of something similar that is encountered in laboratory work on almost a daily basis: Say two types of measurements are made on four samples. A diagram of the data gathered might look like Figure 5-2A. How should such data be interpreted if the two axes reflect important characteristics about the sample and are related to each other?

One interpretation, perhaps the simplest, is shown in Figure 5-2B. But logically, we cannot be sure this is the "true" relationship. Perhaps we will collect more data and eventually find that the original points are best interpreted as shown in Figure 5-2C. Nevertheless, when we have measurements from only four samples, scientists follow Ockham's advice and choose the simplest interpretation, that is, the interpretation shown in Figure 5-2B.

Notice again that the choice to adopt Ockham's advice is a choice about how scientists proceed. Thus it is not necessarily true that, say, the gravitational constant has not changed at all, one way or the other, since the universe began. That is, scientists assume that physical constants have been the same through time, even though they cannot demonstrate that the universe is organized according to the simplest principles. Simplicity is the methodological assumption of science as a profession.

WE DON'T KNOW WHY SCIENCE IS SO SUCCESSFUL

If the scientific method is rather arbitrary—a notion that may surprise some readers—it is even more surprising to discover that at the foundation of scientific research is a reliance on mathematics, which, as we have seen, is not an empirical branch of knowledge. Why does math "work" in the physical world, and work so well? This basic question has not been answered by philosophers, as they are the first to admit.

As a Princeton mathematician wrote in the middle years of this century, "The miracle of the appropriateness of the language of mathematics for the formulation of the laws of physics is a wonderful gift which we neither understand nor deserve. We should be grateful for it and hope that it will remain valid in future research."[5] And in the words of no less a scientist than Albert Einstein: "The world of our sense experience is comprehensible. The fact that it is comprehensible is a miracle."[6] This is delightfully unusual thinking (note that both great men appealed to what we may assume they intended as the metaphor of miracles).

> There certainly are fundamental questions that baffle us, including why the universe is ordered by mathematically precise natural laws and why the human mind can grasp such laws. At present all scientists can say is that they have made great progress—indeed, almost unimaginable progress—toward understanding the physical universe, by using combinations of the methodologies advocated by Aristotle, William of Ockham, and others. Moreover, scientists have chosen to rest their work on the foundation of mathematics, and for reasons we cannot explain, this has proved to be a remarkably helpful choice.

GEOLOGY'S CORE METHODOLOGICAL PROBLEM: THE EVOLUTION OF UNIFORMITARIANISM

Geologists base many of their assertions about the earth's history on the principle they still, perhaps unfortunately, call uniformitarianism. Geologists' understanding of the principle has changed in the last two centuries. Uniformitarianism is now best understood as a methodological assumption, not a substantive claim about the earth's history. From this perspective, the founding principle of geology indicates that it is simply a geologic version of the assumption of the physical sciences we discussed earlier.

[5] Eugene P. Wigner, "The Unreasonable Effectiveness of Mathematics in the Natural Sciences," *Communications on Pure and Applied Mathematics* 13 (1960):14.

[6] David Lindley, *The End of Physics: The Myth of a Unified Theory* (New York: HarperCollins, 1993), p. 4.

Uniformitarianism should not be regarded as a description of how the earth works but, rather, as an important methodological assumption that earth scientists use in their work.

ERRORS ALL AROUND US

An article by James Shea in the early 1980s sketched a dozen common ways that geologists misuse the term *uniformitarian*.[7] The article is instructive, for geologists continue to misuse the word and abuse the underlying concept. In addition, Stephen Jay Gould wrote a more somber sketch of some of the same issues.[8] In the challenging area of methodology, all geologists have made mistakes. Indeed, Shea concludes his otherwise excellent article with an imprecise statement of his own. He writes (as would most other geologists), "In developing and choosing among hypotheses as to the processes and conditions that produce a certain result, as scientists we *must* follow the rule of simplicity [that is, the Principle of Parsimony]."[9]

This statement is true as long as scientists agree to prefer the simplest hypothesis. But scientists' professional preference may have nothing to do with the way the world is actually organized.[10] Therefore, someday scientists may find another, and better, method of choosing among competing and fully adequate hypotheses. After all, what scientists hope matters most are the empirical data of the universe, not the set of procedures they happen to have adopted for their professional work.

On the other hand, gradualism (often incorrectly called "uniformitarianism") has proved to be a useful tool in both the study of the earth's surface and metamorphism and plate tectonics. But gradualism is a substantive claim about the earth, and geologists now believe that the claim is false in many instances. Examples of earth processes that are not gradualistic include catastrophic outburst flooding (which we will address in Chapter 6), the genesis of life on earth (Chapter 8), and episodes of mass extinction (Chapter 11).

[7]James Shea, "Twelve Fallacies of Uniformitarianism," *Geology* 10 (1982): 455–460.

[8]Stephen Jay Gould, "Toward the Vindication of Punctuation Change," in *Catastrophes and Earth History*, ed. W. A. Berggren and John Van Covering (Princeton, NJ: Princeton University Press, 1984), pp.74–99.

[9]Shea, "Twelve Fallacies of Uniformitarianism," p. 4 (italics added).

[10]Ockham's razor and the scientific preference for elegantly simple mathematics kept scientists from appreciating fractals, unstable equilibrium, and the so-called butterfly effects. Reseachers avoided the entire world of chaos theory not because it is not an important part of the universe but because it did not accommodate many of science's methodological preferences.

In practice, geologists often use a three-step approach to constructing explanations that impinge on concepts of uniformitarianism:

1. They start by appealing to geologic processes as they operate today, for example, erosion in the spring, beach migration during storms, or meteorites falling to the earth as small stones.
2. If such explanations fail to account for what geologists see in the rock record, they still invoke the same processes but assign them rates or magnitudes never seen in the present. Examples of this approach are the explanation of the Channeled Scablands, scoured by massive outburst flooding, and enormous craters believed to have been formed by the impact of large meteorites.
3. If geologists still cannot explain what they see in the geologic record, they will invoke processes different from, but related to, what they have seen in their own lives. For example, geologists and other scientists postulate that sometime during the earth's early history, life was spontaneously generated from inorganic materials.

SUMMARY AND CONCLUSIONS

Scientists use a set of methods, collectively termed *methodology*, to guide their work. Geologists and geology textbooks have struggled over how to present concepts of methodology, especially the idea of uniformitarianism. The modern understanding of this term is that it represents assumptions about the similarity of past and present earth processes. Uniformitarianism is not a substantive claim about earth history but instead is an important way of describing how geologists like to think about the earth. Although gradualism is an easy complement to uniformitarianism, there now is room in geological science for catastrophic as well as gradual change. The retreat of gradualism in the face of evidence for sudden change will be a theme of coming chapters in this book.

QUESTIONS FOR DEBATE AND REVIEW

1. Compare and contrast gradualism and catastrophism. Which view more naturally fits with a "young earth" (for example, an earth formed in 4004 B.C.E., as Bishop Ussher proposed)?
2. Were all catastrophists biblical literalists? Are today's biblical literalists (for example, those who subscribe to "creation science") catastrophists?

---

3. What is meant by methodology? Name other disciplines (aside from science) that use particular methodologies.
4. Explain the principle of parsimony and note the nickname by which scientists generally refer to the idea. Is it surprising that a medieval theologian would have contributed such a central idea of scientific methodology? Why or why not?
5. How do geologists evaluate ideas with respect to occurrences in the present, the near past, and the never-seen distant past?

Chapter 6

Catastrophism Revived: The Example of Pleistocene Outburst Flooding

In the eighteenth and nineteenth centuries, leading geologists worked dilgently to circumvent the catastrophic approach to earth history, which was increasingly regarded as secular heresy, a throwback to prescientific thought. James Hutton, John Playfair, and Charles Lyell exploited the benefits of gradualism, as can be seen in the title of Lyell's 1872 publication: *Prejudices Which Have Retarded the Progress of Geology*. The chief and most destructive prejudice was, of course, catastrophism. Lyell and his followers were combating what they saw as a reactionary view of earth processes.

By the early twentieth century, gradualism had triumphed; it was now assumed to be the fundamental characteristic of our planet's history. But a practical shortcoming of gradualism had been noticed by many geologists, among them an American, J Harlen Bretz of the University of Chicago, who worked to show one instance of catastrophic change to the entire geologic community. We will focus on Bretz's work because the subject is an easy one for a nonspecialist to understand. Bretz argued—almost alone and for several decades—that barren rocky fields of unusual topography in eastern Washington State could not have been formed in the way that the gradualists claimed. Rather, he maintained, the topography must have originated by catastrophic flooding on a scale that geologists had not imagined since the days in which Noah's flood was taken seriously in the scientific community.

The Unusual Geology of Washington State

A WEIRD AND BARREN PLACE

The Channeled Scablands of central and eastern Washington State show unusual surface features, including long and deep channels, which local residents call coulees (from the French "to flow"). The coulees cut down through the loess (loess is fine-grained soil blown into dune shapes by the wind (Figure 6-1); the name comes from the German word for "loose") and slice into tough, dark, basalt rock. As seen from the air, the coulees form a complex braided pattern (Figure 6-3). Grand Coulee is a well-known and huge channel in the Scablands, now the site of one of the world's largest hydroelectric dams.

The Channeled Scablands are easily studied by geologists, for they are almost exclusively barren rock and gravel (Figure 6-4). Vegetation is sparse because there is little or no soil. The rich loess laid down elsewhere in eastern Washington during the Pleistocene (the Ice Age) has been stripped away from the aptly named Scablands.

In the early part of this century, geologists entertained two ideas about how the coulees of the Scablands could have formed. First, some believed

FIGURE 6-1: These loess hills are shaped like sand dunes. They were formed from windblown dust that accumulated in dunes during the Pleistocene (the last 2 million years of earth history). The hills cover much older basalt rock. (Photo: E. K. Peters.)

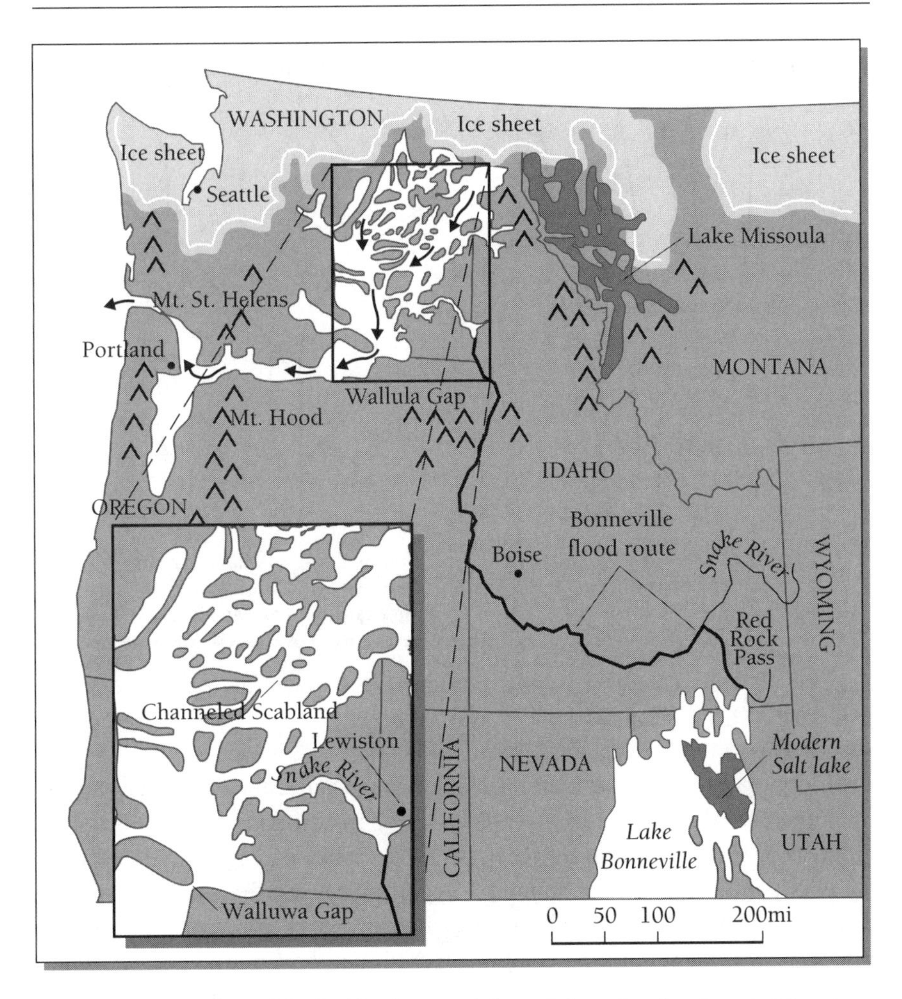

FIGURE 6-2: Coulees of the Scablands, carved through loess into basalt rock. Map view as seen from above.

the channels had been cut by glaciers during the Pleistocene. It was well established that during the 2 million years of the Pleistocene, an enormous continental ice sheet had advanced from Canada into the northern United States and then retreated. A total of four cycles of advances and retreats had been documented in the Midwest. The coulees of Washington State are large and broadly U-shaped in cross section, in accord with their formation by glaciers. And glaciers were clearly active elsewhere in Washington, for example, around Puget Sound and in the Cascade Mountains. Giant granitic boulders can be found in the Scablands, separated by a hundred miles or more from

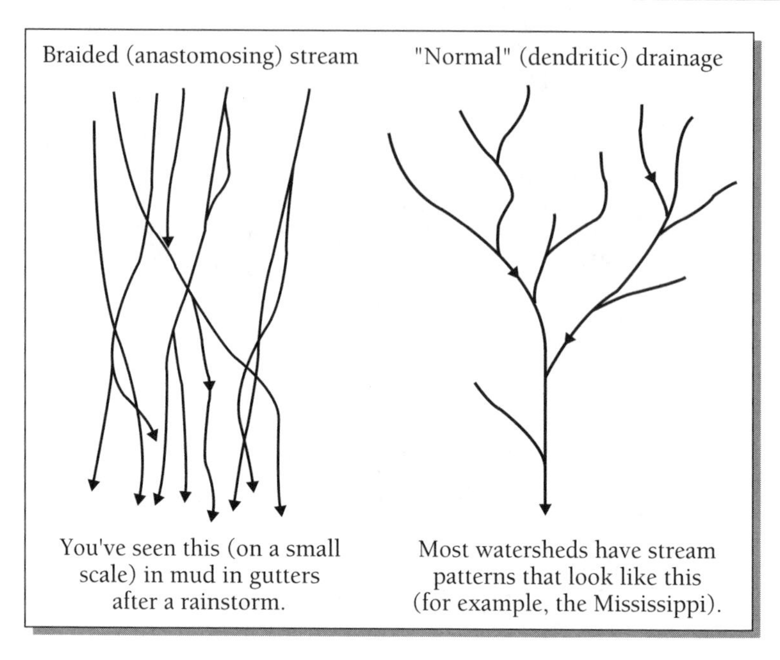

FIGURE 6-3: Different types of stream drainage. The braided stream pattern is what we see in the Scablands.

their origin in Montana or Idaho. Geologists who subscribed to the glacial theory interpreted the boulders as examples of the familiar concept of glacial erratics, or rocks carried in the ice of large glaciers a long way from their origin.

A second and quite different hypothesis was that the Channeled Scablands had been carved out by fluvial (river) processes during the Pleistocene. This school of geologists argued that the Columbia River, which is large even today, would have been much larger during the ice ages when it was draining meltwater from the extensive Canadian glaciers. A much larger Columbia might have spread out during floods and carved the coulees. Then, as the water level dropped and the Columbia River became smaller, the river would have retreated to its present channel, leaving the Scablands high and dry.

Both these two views—the glacial theory and the Columbia River hypothesis—fall within the framework of gradualistic uniformitarianism. That is, both appeal to processes that geologists can see and study at work today. There are glaciers grinding away rock in alpine valleys in Washington at this moment. Geologists do not appeal to anything fundamentally new if they say that glaciers cut across Washington during the Pleistocene and carved the coulees. Similarly, the Columbia River is large, and it certainly could have been

Figure 6-4: This coulee, the Upper Moses Coulee, is characteristic of the Scablands. (Photo: Copyright John S. Shelton.)

larger in the past. The normal flooding process of rivers can be studied today and perhaps can account for the Scabland's channels, all of which are indeed near the Columbia.

In essence, both these theories allowed geologists to study the Scablands and link their work to processes still operating. This was the type of geologic research favored in the early part of this century. Geologists had been led by Lyell and other great men to a simplistic kind of uniformitarianism. But despite the methodological assumptions of most geologists, it was in fact possible that the Channeled Scablands of Washington State were an example of an event in the geologic past that could not be studied in terms of currently active processes, at least on anything like the relevant scale.

Bretz Adopts Quite Another View

J Harlen Bretz was fascinated by the unusual features of the Channeled Scablands and spent many summers in the 1920s and 1930s walking through the coulees with a mule. His notes do not say whether the mule was a good companion, but the time Bretz spent in the field was intellectually fertile. Summer after summer, he returned and took walking tours of both the Scablands and the land downstream. His observations were not complex; he took no esoteric measurements. But Bretz was an original thinker.

Field evidence made Bretz sure that the coulees had not been carved by glaciers. There was no glacial till (Figure 6-5) in central Washington, and

FIGURE 6-5: Glacial till on the side of a moraine. Note the lack of sorting by size (both large and small particles are mixed together—-the largest is 1 yard across). Note also the angularity of most particles. Both these features are characteristic of debris moved by glaciers. (Photo: E. K. Peters.)

no moraines could be found in the Scablands. He also rejected the fluvial theory of flooding, believing that the morphology of the Scablands was not formed by a river, not even a giant Columbia River at flood stage.

In 1923 Bretz published his first paper on the Scablands, a descriptive piece of work summarizing his field observations. He omitted any explanation of the Scablands' origin, to avoid attacking the prevailing dogma of uniformitarian gradualism. Bretz's article did, however, begin to publicize the area's unusual topographic features. Shortly after publishing this descriptive article, Bretz found the courage to put forward a truly catastrophic hypothesis to explain the origin of the Scablands. He wrote that a wall of water, 1000 feet high, had slashed through the area, cutting down through the loess and the basalt. He gave as evidence of this massive outburst flooding the following:

1. The crisscrossing complex of the channels: The Scablands look like a gigantic braided stream, a channel pattern that indicates rapid inundation by water followed by its rapid retreat. Furthermore, the bottom of many coulees have no streams at all running in them, and some of them are extraordinarily flat, without the hint of a V shape. In fact, this unusual shape lies behind the important words *coulee* or *channel* rather than *valley*. There are no normal valleys throughout the Scablands.

2. Deep gravel bars in the middle of the channels and at the perimeter of the area. Some of the gravel bars in the coulees are 100 feet high and could have been formed only by floodwaters much deeper than that. One bar along the eastern boundary of the Scablands rises impressively at the edge of the large loess hills. The well-sorted basalt gravels in this bar make it clear the huge volume of material cannot be interpreted as moraine. The bar looks like a waterborne deposit, but it is more than 100 miles from the Columbia. This enormous bar stretches along the edge of the Scablands for more than 10 miles.

3. Cataract cliffs and plunge pools hundreds of feet in diameter. Such cliffs and pools are now completely dry or occupied by only small amounts of water, which seem unable to have formed such massive features. Two fine examples are Dry Falls, near Grand Coulee (shown on the cover of this book), and Palouse River Falls, near the Snake River, shown in Figure 6-6.

4. Thick strata of fine silts upstream of the Columbia and Snake Rivers and in their tributaries. Bretz believed that the tributaries of the

Figure 6-6: Palouse Falls. The huge basalt rock amphitheater seems to indicate that in the past this coulee contained much, much more water. (Photo: E. K. Peters.)

area's two major streams must have had to flow backward when a wall of water hundreds of feet high reached them. This backflooding produced the fine sediments preserved in tributary valleys. Beds of fine silt with isolated ice-drop cobbles can be found up the Snake River Canyon as far away as Idaho.

5. Giant ridges, up to 50 feet high, in the gravel bars in the coulees. Interpreting these ridges as normal ripple marks on a gigantic scale, Bretz argued that the coulees must have been filled with water hundreds of feet high and moving in excess of 50 miles per hour (Figure 6-7).

6. Longitudinal bars, cutting across hanging valleys. These bars were laid down at the edge of the floodwaters, sometimes forming natural levees and cutting across the mouths of what later became hanging valleys. Lakes formed in these valleys after the flood subsided, since normal drainage was no longer possible.

7. Chaotic or deranged drainage patterns over much of the area. Normal dendritic drainage is not found in the Scablands, which are

FIGURE 6-7: Giant ripples seen from the air. (Photo: Copyright John S. Shelton.)

dominated by closed depressions and coulees. This indicates that the currently active small streams did not form the topography.

8. Smooth, bowl-like hollows in the basalt. These features, carved in solid rock, look like plucked and scoured beds (see cover photo).

THE AUTHORITIES RESPOND

Bretz's bold papers attracted attention, distressing the leading geologic authorities of the day. Bretz, they feared, was reverting to a catastrophic scheme that belonged in the profession's distant past. Geology had progressed for more than a century on the basis of a gradualistic view of uniformitarianism, so it was not unreasonable for thoughtful geologists to dread any backsliding into catastrophism. On what, they asked, might Bretz focus next? To gradualists, there seemed no limits to catastrophic thinking.

In 1927 many distinguished geologists gathered in Washington, D.C., for a professional meeting. Although Bretz was invited to defend his views, the invitation appears to have been intended as an opportunity to criticize his arguments. Accordingly, Bretz was attacked in the name of uniformitarian gradualism, and his concepts about massive flooding were denounced. Even though most of Bretz's critics had not seen the Channeled Scablands, they were sure, as a matter of principle, that the forces acting in eastern Washington State could not have been catastrophic.

> Bretz's critics' reluctance to accept Bretz's idea was not terribly narrow-minded or stubborn, for gradualism—and interpretations flowing from it—had been extremely helpful in many geologic problems. Now, however, Bretz was asking people to think quite differently, and unless he had compelling evidence, other geologists were understandably reluctant to give up a style of interpretation that had worked well in the past. Nevertheless, from Bretz's point of view, the authorities were confusing methodology with substance.

In retrospect, Bretz's critics did have one valid point. Where did the water for this unprecedented flooding come from? Bretz had apparently not focused on this problem, and it stood as the cornerstone of reasonable challenges to his views. Today, one might look back at the youthful Bretz and say, It is nice to be right, but in science it helps to be right for the right reasons and to explain those reasons to all who ask. A source of the enormous quantities of water that Bretz was relying on was a reasonable "next question" to address.

THE GENERAL DEBATE ROLLS ON

Another point of interest about the 1927 meeting is a storyteller's delight. Apparently, a young geologist named J. T. Pardee was sitting in the audience, listening to Bretz's talk. When the respected authorities criticized Bretz for failing to offer a source for the enormous quantities of water that his theory demanded, Pardee said to a friend: "I know where Bretz's flood came from; it came from [Pleistocene] Lake Missoula in western Montana." But as the story goes, Pardee knew how to get along in the professional world, and at this point, his idea may only have been a hunch. Since those who made decisions about hiring and promotion all were highly skeptical of Bretz's ideas, Pardee wisely decided not to speak out at the meeting. He did, we think, speak privately to Bretz, but Pardee's public silence is a clear example of the tendency of many scientists not to publish or discuss data, and especially casual hypotheses, that may seem outrageous to their colleagues. All professional disciplines, of course, have this inherent conservatism. For his part, Bretz may not have advanced Pardee's idea either because he felt he could not prove that his floodwaters came from Montana or because, as a matter of professional courtesy, he would not "steal" another geologist's ideas.

Throughout the 1930s, Bretz continued his publications on the Scablands, amassing a longer and richer descriptive data base for the area.[1] But out of stubbornness, arrogance, or fear, Bretz largely ignored the question of the source of the floodwaters, and this gap in his views meant that his position did not attract many supporters. Eventually, several of the important and "gradualist" geologists of the day visited the Scablands in a series of field trips, but they concentrated on modifying the glacial and fluvial hypotheses. Although they tried to take into account some of what Bretz had described, they did so in a framework of gradualism. These

[1]Here we see that scientific culture does have a significantly objective dimension. Even though Bretz was out of favor with the most prestigious geologists of his generation, his work continued to be published in the best geologic journals. Some historians of science would say this is noticably different from what happens in ideologically driven fields (such as politics and economics), in which people marginalized by the major authorities have a much more difficult time getting their views published in the field's principal journals.

revisionist theories were respected by many geologists around the country and were taught to students. But everything changed when Pardee finally spoke up.

Pardee Finds His Voice

In 1942 Pardee published a description of enormous ripple marks on the floor of the Pleistocene Lake Missoula. Pardee had found good data to back up the hunch he had had at the meeting long before.

The "ripples" that Pardee described were gravel bars 50 feet high and 500 feet apart. Indeed, they had formed on such a large scale that they had previously been mistaken for normal hills. J Harlan Bretz himself had not understood them when he visited the area in the 1930s. But photography from the air was becoming increasingly common, and Pardee's paper showed that the landforms were just like familiar ripple marks, only expanded to a scale no one had thought to consider. Pardee also described eddy deposits composed of gravel that had formed in the relatively slack floodwater behind promontories. Pardee offered a catastrophic idea—an ice dam had formed the huge lake (Figure 6-8), and when the ice broke, enormous quantities of water moved down the Clark Fork valley toward Spokane, Washington. The "mega" ripple marks were

Figure 6-8: The ancient shorelines of Lake Missoula can be seen on the huge hills behind the city of Missoula, Montana. (Photo: Copyright John S. Shelton.)

formed at the start of a flood of what one might call biblical proportions. Pardee made a good case for a huge source of water that had drained very rapidly to the west, thereby vindicating J Harlan Bretz.

BRETZ'S VIEWS ARE ACCEPTED AND REFINED

Both Bretz and others began to recognize different episodes of massive, outburst flooding in the Scablands. The largest flood may have come first, but there were dozens of episodes of catastrophic flooding as the ice dam in Montana reformed when glaciers from the north advanced, and it collapsed when the glaciers retreated or the ice broke of its own accord. It was shown that there were 35 to 40 rhythmites (Figure 6-9) in Lake Missoula sediments, that is, graded beds with very coarse material on the bottom. In between the rhythmites were varves (a pair of layers of alternately finer and coarser silt or clay), each one believed to be the sedimentary record of annual precipitation changes. On average, there were about 50 varves (corresponding to 50 years) between the rhythmites (the evidence of massive floods).

In the 1950s a number of academic geologists took field trips to the Scablands. This time the trips led to a series of publications proposing repeated and massive Bretz-like floods. Rhythmites were found all over the area. In time, it became common to argue there had been as many as 70 such episodes of outburst flooding, with the first being the biggest and the dirtiest.

But catastrophism still stuck in the throat of some members of the established geologic hierarchy. Only in 1971 did the last anti-Bretz gradualist, a professor at Yale, concede the point of massive flooding. As Bernard Barber, Thomas Kuhn, and others have repeatedly pointed out, scientists do occasionally resist good science. Still, it is important to note that empirical evidence did overcome the methodological stubbornness present at even Yale.

In 1979 Bretz won the Penrose Medal of the Geological Society of America, the nation's highest geological award. In the end, the "truth" of catastrophism did win, but it took more than 50 years to see the project through. And this was despite the fact that the event in question was relatively easy to document, through field observations requiring only a notebook and a good mule.

In fairness to Bretz's early critics, however, the outburst flooding he called for was so great that it is difficult to visualize even now. Geologists believe that for 2000 years in the late Pleistocene, about 40 outburst floods cut through what is now northern Idaho and eastern Washington. The Snake

FIGURE 6-9: Rhythmites in Scabland sediments. Coarse rock cobbles are capped by finer and finer sediments. Two and a half cycles are shown (see the drinking bottle for scale). (Photo: E. K. Peters.)

FIGURE 6-10: This view from the air shows "islands" of soft loess surrounded by flood-ravaged basalts of the Scablands. (Photo: John Shelton.)

River ran backward when the wall of water reached the Columbia. The Columbia Gorge, between Oregon and Washington, was scoured deeply by the cataclysmic floodwaters. All this resulted from repeated advances of glacial ice in western Montana, ice that dammed the Clark Fork River and formed a huge, deep lake. Bursting ice dams released water sufficient to cover about 15,000 square miles of land at a depth of several hundred feet (Figure 6-10).

HUMAN HISTORY AND THE FLOODS

You may wonder whether there were human witnesses to the floods that Bretz documented. There is good reason to believe that native peoples saw, and quite possibly perished in, the violent floods of water, ice, and rock that swept across the Columbia Plateau. It has been estimated that the wall of water moved at about 50 miles an hour, leaving no hope that humans, or any other wingless animals, could have escaped. A blast of compressed air would have preceded the wall of rapidly moving water, and although the roar it produced may have been heard for half an hour before the water arrived, it is unlikely anyone could have known what the sound was.

Human populations must have been repeatedly decimated by the events forming the Scablands. But the last of the floods occurred long ago, by human standards, and the native peoples offer no stories that unambiguously refer to the magnitude and violence of the Bretz flooding.

Catastrophic Floods Pop Up Everywhere

Another Part of the Western United States

After Pardee's publication helped confirm Bretz's hypothesis, other geologists began to look at their own field areas with an eye slightly more sympathetic to catastrophic rates of change. The clearest evidence for a different, but equally enormous, Pleistocene flood came to light in Utah and Idaho. In the nineteenth century, the well-known and respected geologist G. K. Gilbert had described the huge extent of the Pleistocene Lake Bonneville, centered on the current Salt Lake in Utah, but much larger (Figure 6-2). The Salt Lake basin is closed (that is, it has no outlet). In the wetter and colder climate of the Pleistocene, Lake Bonneville rose higher and higher. Gilbert proposed that Red Rock Pass in southern Idaho had become an outlet for the lake when it rose to the level of the pass. Lake waters, Gilbert believed, carved erosional features into the rocks of the pass as they cascaded down toward the Snake River drainage system and rapidly cut down into the pass in the process. Once the pass was breached in this manner, the downcutting would have become even more rapid. This would have quickly lowered the level of Lake Bonneville, and water pouring over the pass would have led to further, rapid erosion of the lake's rim. Gilbert thought that a flood on the order of that of the Missouri River was indicated by the evidence, certainly a major event but hardly dramatic by catastrophic standards.

In the 1950s, however, evidence connected with the sudden drainage of Lake Bonneville was reexamined in light of Bretz's work. Geologists began to wonder whether the huge gravel bars and boulder deposits on the Snake River Plain and in the side canyons of Hells Canyon were remnants of the

(A)

(B)

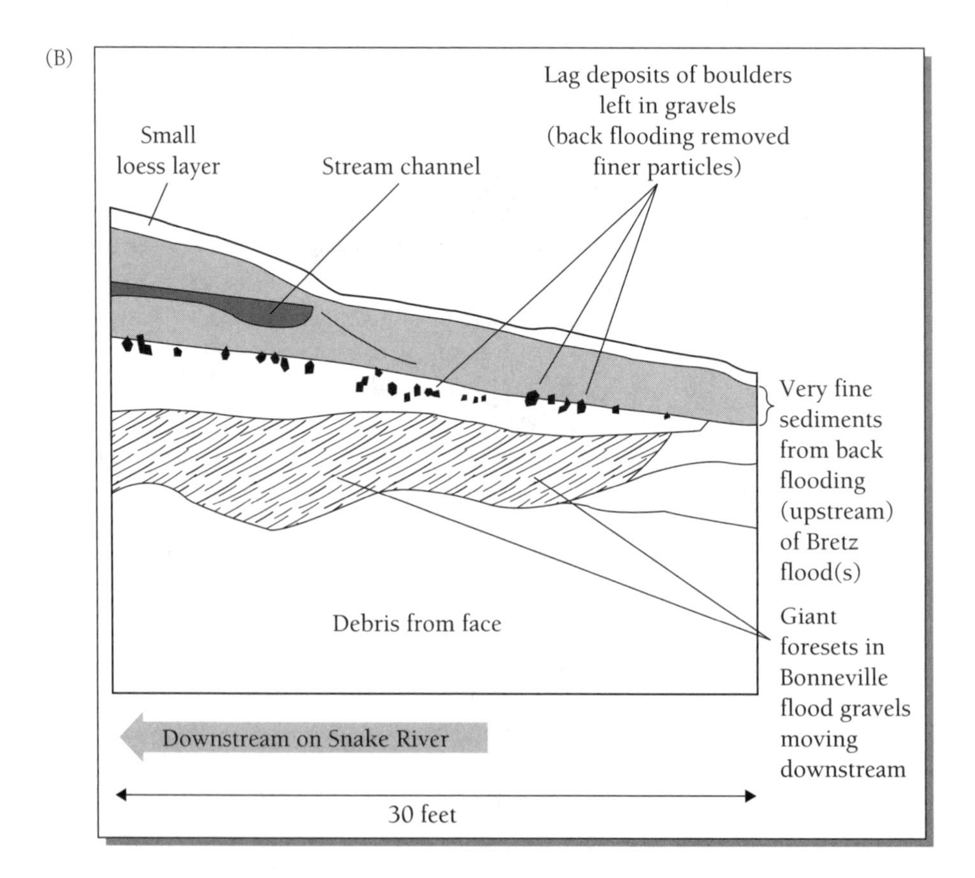
Small loess layer
Stream channel
Lag deposits of boulders left in gravels (back flooding removed finer particles)
Very fine sediments from back flooding (upstream) of Bretz flood(s)
Giant foresets in Bonneville flood gravels moving downstream
Debris from face
Downstream on Snake River
30 feet

FIGURE 6-11 (FACING PAGE): Outcrop in Lewiston, Idaho, on the Snake River, shows the Bonneville flood gravels (with megaforeset beds) overlain by Bretz's back-flooding deposits. (Photo: E. K. Peters.)

Bonneville flood. The scale of the flood, it became clear, had been enormous. Once again, morphologic features (the shape of the landforms) had not been recognized, simply because of their size.

Today, gravel bars hundreds of feet thick that were created by the Bonneville flood have been recognized as far downriver as Lewiston, Idaho, a town on the border of Idaho and Washington. The Bonneville flood event is now dated at approximately 14,500 years ago. It is believed to have discharged 4750 cubic kilometers of water into the Snake River drainage system. Bretz's work helped clear the way for a new understanding of southern Idaho's Pleistocene history because he had almost single-handedly made catastrophic flooding an acceptable geologic hypothesis.

As it happens, a gravel pit just outside Lewiston's city limits gives an excellent exposure of Bonneville gravels, with their foresets pointing down the Snake River (Figure 6-11). These gravels are overlain by back-flooding Missoula beds, which have foresets going up the canyon. Bretz's large sense of vision is needed to imagine two catastrophic floods coming from different directions, each inundating the Snake River floodplain in a manner almost more than even biblical descriptions can capture.

EVIDENCE FROM OTHER PLANETS

In a final vindication of Bretz's hypothesis, more exquisite than anything he may have dreamed, NASA probes imaged the surface of Mars in the 1970s. They found huge channels and a morphology similar to those of the Channeled Scablands, but on an even larger scale. It is now clear that J Harlan Bretz's catastrophic hypothesis has been confirmed quite out of this world.

SUMMARY AND CONCLUSIONS

Gradualism and catastrophism have clashed repeatedly in the history of geologic inquiry. When this century began, gradualism was the established and orthodox framework of thought for geologic research. It therefore took a long time for J Harlen Bretz's catastrophic interpretation of the Channeled Scablands to get a fair hearing. If he had not been persistent, the catastrophic flood theory would not have been accepted as early as it was. But once geologists had learned the lessons of the Channeled Scablands, they reexamined other field evidence in the western United States. Ultimately, even the morphology of another planet has seemed to confirm Bretz's vision of outburst flooding.

Bretz's triumph does not mean that gradualism is not the best approach geologists can use to investigate many geologic problems. Rather, scientists must keep in mind that their methodological assumptions are merely descriptions of how geologists like to think, not descriptions of earth history itself.

The moral of the Bretz story is that geologists must be very careful not to confuse method with substance, a distinction crucial to doing scientific research. It also is essential that nonscientists keep this difference in mind when evaluating what scientists can be expected to do in addressing our society's many challenges, a topic to which we shall return.

QUESTIONS FOR DEBATE AND REVIEW

1. If aerial photographs of the Scablands had been available in the 1920s and 1930s, Bretz's theory might have been accepted more quickly. The scale of the Scablands' features is best seen from the air; even Bretz himself walked over Pardee's ripple marks but did not see them for what they were. What other examples do you know that connect technological development and scientific discovery?
2. What personal and professional qualities made Bretz successful? Is it reasonable to expect that all scientists share those characteristics?
3. Imagine how confusing catastrophic outburst flooding must have been for the (few) humans who survived one of the floods. How do you think such survivors might have described their experiences? How might they have tried to explain what had happened?
4. What other massive floods were better understood in light of Bretz's work in the Scablands?

Chapter 7

Gradualism Strikes Back: Drifting Continents and Plate Tectonics

Even while J Harlan Bretz toiled to make geologists aware of catastrophic changes, other researchers struggled to understand the deformation of the earth's crust in gradualistic terms. For several centuries, geologists had tried to explain the origin of mountain chains. Metamorphic rocks in the core of mountain belts clearly reflected a long acquaintance with great heat and pressure. But geologists could not understand the causes behind enormous mountains like the Andes, Rockies, and Himalayas.

Progress in understanding the deformation of the earth's crust came from two quite distinct sources: In 1915, a German scientist named Alfred Wegener offered a theory of mobile continents, and in the 1950s and 1960s geophysicists explored the ocean floor. The story of how these two separate efforts combined to form the modern theory of plate tectonics is the subject of this chapter.

Deforming the Earth's Crust

We Have Done Some Things Right!

In this book, we have emphasized the human nature of all scientific research and the aspects of geological science that make some of its theories impossible to test directly and experimentally. But geology is, nevertheless, an intellectually productive discipline. In this century, geologists have established the overarching theory of **plate tectonics,** a framework that has been phenomenally successful in interpreting many different kinds of data and observations. Scientists place special value on broad theories, such as plate tectonics, that support other, smaller theories. The path that geologists followed to create plate tectonics contains several lessons about how geological science progresses.

The concept of scale is important to remember when reviewing scientific work, especially because it is almost never explicitly addressed but is generally implicitly assumed. Scale is crucial; it can guarantee the omission of some features and therefore the failure to develop certain lines of inquiry. As a matter of convenience, early geologists studied hand samples of rocks, that is, pieces that can be hammered off a rock outcrop, held in the hand, and carried back home. A great deal of time in traditional geology courses is still spent teaching students to recognize and interpret features visible on the hand-sample scale. Small-scale features of rocks and individual mineral grains began to be studied at the end of the last century with the newly invented polarized light microscope and are now studied on an even smaller scale using electron microscopes. But some features of the earth, such as the Hawaiian volcanic chain or the San Andreas Fault, demand study on a vastly larger scale.

One indicator of the power of tectonic theory is that it fits at least some of the data that geologists have gathered on the microscopic scale, the hand-sample scale, and the scale of observation visible to astronauts in space. But the theory would never have arisen from only small-scale observations; it required attention to a whole-earth framework to be fully developed.

WHY DO MOUNTAINS RISE ABOVE THE CONTINENTAL PLAINS?

Early geologists had only hazy ideas about the origin of the earth's numerous mountain ranges (Figure 7-1). To many, it seemed clear that forces of enormous magnitude had caused upheavals of the crust that were driven by hot gases escaping from the interior of the planet. The earth, it was assumed, was contracting as it cooled, and these contractions were shrinking the surface of the entire globe. This theory had the endorsement of early physicists.

In the nineteenth century, geologists who specialized in mineralogy formulated laws of complex crystal symmetry. Soon some geologists speculated that the entire earth was a kind of crystal, growing smaller as it cooled. Mountain ranges, it was thought, must be arranged with crystalline symmetry on the surface of the planet. Elie de Beaumont (1798–1894), working in France, spent a great deal of time trying to match pentagonal patterns on the geologic map of

FIGURE 7-1: The height and depth of even modest mountain ranges, such as those shown here in northeastern Oregon, indicate massive forces at work. But how these mountains could have been lifted above their surroundings is not obvious to most of us. (Photo: E. K. Peters.)

Europe. His geometric drawings became more and more complex as he made more and more observations about European mountains and rocks.

The noted American geologist J. D. Dana (1813–1895) accepted the notion of a buckling earth's surface but abandoned ideas about geometric regularity. He pointed out that continents are mostly granite and stand high above the basaltic ocean basins. Dana accepted the theory of **isostasy,** which states that less dense, but thicker, rock—for example, granite—stands higher than denser rocks do because both are "floating" on a semiliquid layer beneath the crust (Figure 7-2).

Dana argued that great lateral pressures in the crust made some zones of the lighter granitic rock move up and down. This vertical movement, located in zones he called **geosynclines,** gradually eased lateral pressures and slowly gave rise to mountains on the continents. Dana and other geologists who accepted a substantial vertical movement of the continental crust were certain that the continents did not move laterally. Such movement would require that the granitic rocks somehow plow through the solid basalt of the ocean floor. Nothing, it was said, could account for anything so unlikely.

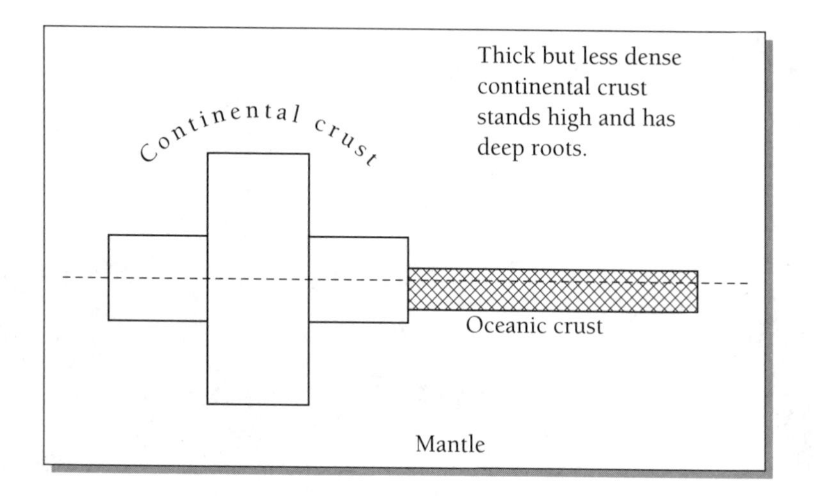

FIGURE 7-2: Isostasy. The thick but less dense continental crust stands high and has deep roots.

What might account for such movement, of course, would necessarily lie deep within the earth. Dana, like most other geologists of his day, assumed that the proper domain of geology was the earth as it could be observed by geologists. That is, geologists should investigate the outcrops and hand samples of rocks available at and near the surface of the continents. Speculation about what was deep inside the planet departed from the certainty of scientific observation, these geologists thought, and led into fantasy.

The theory of geosynclines was accepted as the best framework for geologic research, and it was taught to students through the 1960s. But from the turn of the century onward, evidence had been mounting from a variety of disciplines in earth science that the continental crust moved laterally and that the oceanic crust was quite unlike what Dana had assumed. A complete theory giving scope to mobile continents was developed in the 1960s and was rapidly embraced by virtually all professional geologists.

But the sweeping change in the overarching theory concerning the earth's crust did little to alter the detailed research then under way. In time, however, scientists incorporated plate tectonics into every subfield of geology. Nevertheless, most specialists still concentrate on questions that would have interested them before the plate tectonic framework crushed earlier ideas about the structure of the earth and the distribution and origin of mountain ranges.[1]

The Fitful Growth of a Theory

Nonscientists of the nineteenth century had occasionally noted the complementary similarity of continental coastlines, notably those of South America and Africa. But a theory of lateral continental movement awaited a German trained in meteorology, Alfred Wegener (1880–1930). Wegener did more than look at the shape of coastlines. He noted that those geologists subscribing to Dana's view and to isostasy theory had already postulated a liquid, albeit viscous, beneath the crust, on which granitic rocks moved vertically. There was no reason, Wegener argued, that continental rocks could not move laterally, given a sufficient sideways force. He assumed such movement was at least fairly gradual and cited increasing evidence from geologists working in the Alps that showed the extensive lateral displacement of rocks along thrust faults and giant folds. These features were not in accord with the theory of vertically moving geosynclines, but they fit well with Wegener's ideas of lateral deformation of the earth.

Paleontologists had gathered evidence of similarities between ancient fauna in the Mesozoic rocks of Africa and South America. Wegener emphasized that certain rocks, as well as fossils, were similar on the two sides of the Atlantic Ocean. Indeed, the deepest (oldest) strata of westernmost Africa and easternmost South America were much alike. This similarity continued upward to a point of divergence—the time, Wegener believed, when the southern Atlantic Ocean began to form between the two continents, splitting them apart.

Wegener argued that all modern continents had once been grouped together in a single landmass, which he called **Pangaea.** The southern part

[1]For a discussion of the difference between geologists and those who work in the newer and more prestigious discipline called earth sciences, see the entertaining history written by Robert M. Wood, *The Dark Side of the Earth* (London: Allen & Unwin, 1985). Wood—perhaps not an unbiased observer—contends that another way of viewing the history of continental drift and tectonic theories is that descriptive, indeed almost unscientific, geologists were overtaken by the rigorous work of applied physicists. In recognition of their intellectual triumph in this regard, geophysicists transformed geology departments across this country into earth science departments and brought the study of the earth out of its long, dark ages dominated by descriptive fieldwork and detective storytelling. This small book that you are reading, however, is written from the perspective of traditional geology, for that is the body of knowledge encountered by undergraduates in introductory geology courses.

of the Atlantic Ocean was born in the late Mesozoic, Wegener asserted, when Africa and South America began to drift away from each other. Wegener reasoned that Europe and North America drifted apart in a similar fashion much later, perhaps during the last Ice Age. Wegener's theory was termed **continental drift.**

RESPONSE AND DEBATE

Wegener's hypothesis was initially considered seriously in Germany, but it soon was set aside. "Drifters" were numerous, however, in southern Africa, South America, and Australia. These geologists who worked on the fringes of European culture—in what might be considered frontier conditions—were clearly more receptive to Wegener's idea. Unfortunately, most scientific journals were edited, and prestigious scientific prizes awarded, in Europe, and Wegener cared deeply about the reception of his theory in these established circles.

In England and America, Wegener's views were vigorously attacked. A distinguished mathematician at Cambridge University, Sir Harold Jeffreys (1891–1989) vehemently opposed all ideas related to continental drift. Jeffreys's prestige as a mathematician added weight to his arguments about the geophysical impossibility of moving continents.

Charles Schuchert, a prominent American paleontologist, criticized Wegener's claims about fossil evidence: Although the Mesozoic fauna in South America and Africa are similar, they are not the same. Furthermore, Wegener had been sloppy in his presentation of the fossil evidence, lumping together species from different periods of time.

Paleontologists (geologists who specialize in the study of fossils) were used to explaining what similarities they saw in fauna separated by oceans by postulating ancient, thin land bridges. These bridges, they believed, had later sunk into the oceanic crust and disappeared. The American paleontologist G. G. Simpson had devoted his professional life to the evolution of land mammals. Based on the fossil evidence he had gathered, he believed that the worldwide distribution of mammals and similarities among mammalian species argued against moving continents. He instead emphasized the possibility of animals crossing the sea by chance on natural rafts torn loose by storms. In short, most paleontologists did not endorse drift theory, despite Wegener's claims that the fossil evidence supported his views.

The scientific atmosphere of the 1920s, 1930s, and 1940s was highly anti-Wegenerian; the list of Wegener's supposed deficiencies was a long one. The final entry might be summed up by the question, How could drift theory explain that the alleged Pangaea continent had been stable from the birth of the earth until relatively recent times but then drifted into many separate pieces?

THE OTHER SIDE OF SCIENTISTS

The debate about continental drift sometimes revealed elements in scientists that were a bit less than professional. Proponents and opponents grew exasperated with each other. The American geologist R. T. Chamberlin reflected this impatience when he condemned Wegener and his supporters for their presumed professional shortcomings:

> Wegener's hypothesis . . . is of the footloose type, in that it takes considerable liberty with our globe, and is less bound by restrictions or tied down by awkward, ugly facts than most of its rival theories. His appeal seems to lie in the fact that it plays a game in which there are few restrictive rules and no sharply drawn code of conduct.[2]

It is fair to say that Wegener, on his side, often denounced his critics for their supposed tendency to close their eyes to the data. Although we might in retrospect be sympathetic to Wegener's frustration, it is important to remember that his published work did contain some mistakes. His errors regarding the fossil record in South America and Africa are one example. But geology is a difficult discipline to master because many different lines of evidence are needed to confirm a theory. No one can be an expert on all the subfields that can be called on in a wide-ranging dispute like continental drift. It is not surprising, therefore, that a visionary theorist can make mistakes, even mistakes of breathtaking dimension.

It might be said that during the early decades of this century, continental drift theory generated more heat than light. But most researchers simply continued to do good work on the narrower questions that had always interested them.

[2]A. Hallam, *Great Geological Controversies* (New York: Oxford University Press, 1983), p. 152.

AN ALTERNATIVE TO WEGENER

Since one of the greatest objections to drift theory was the difficulty in moving continents through oceanic rock, a few geologists developed an alternative viewpoint that preserved part of the Wegenerian framework while abandoning other aspects of it.

If the earth were expanding in size and if it were doing so because its ocean basins were growing, it could be that in early Paleozoic times, Pangaea (the landmass that in Wegener's theory incorporated all the continents) almost entirely covered the earth. At that point, there would have been almost no oceans. Then as the earth expanded, the oceans grew, and the giant landmass was split apart. Naturally, fossils and rocks correlate across ocean boundaries. They do not so, however, because the continents are moving but because the ocean crust has grown dramatically owing to the force of the expanding earth.

No powerful ideas were offered to explain why the earth might be expanding, but geologists were tempted to ignore the constraints of physics when geological observations supported a hypothesis. Besides, it could be the case that the force of gravity (the gravitational constant G) was lessening through time for some unknown reason, thus allowing the matter of the earth to shift farther apart. In any event, once the geologists got over the alien idea of a continuously growing planet, this hypothesis allowed parts of Wegener's arguments a new round of consideration. The distinguished geologist Harry Hess invited those propounding an expanding earth to speak at Princeton. Hess's willingness to listen to large-scale hypotheses and his interest in the ocean floor contributed to his success in understanding a better way to use parts of Wegener's theory.

THE INFORMATION GAP: WHAT WAS UNDER THE SEA?

In retrospect, both Wegener's sympathizers and their many opponents lacked key information that would have helped resolve their differences: That is, before the late 1940s, no one knew anything about the geology of the seafloor, simply because the rocks under the ocean were inaccessible to geologists. But after World War II, sophisticated ships that imaged, dredged, and drilled the ocean floor also investigated the geology of the previously neglected parts of the earth's surface. When the topography (shape) of the ocean floor was mapped, two features grabbed geologists' attention. These were high, linear features in the middle of the oceans, areas marked by unusual heat flow and seismicity. Such features are symptoms of volcanic activity. It was known, of course, that Iceland was made of young volcanics, but the data from ocean studies showed that Iceland was only a tiny part of the long Mid Atlantic Ridge and that volcanism stretched in linear swaths around the globe under all the oceans. In addition, scientists mapped deep trenches in other parts of the seafloor. Perhaps most important, as the techniques of radiometric dating of rocks advanced and rock samples from all parts of the ocean floor became available, it became apparent that the entire ocean floor was much younger than most rocks of the continents. Although some continental rocks are several billions of years old, there was nothing at all comparable beneath the seas. That is, the ocean basins were younger than anyone would have guessed.

It is important to note that geologists pursued their work about the ocean basins without special reference to Wegener's theory, relying instead on smaller frameworks of thought. But because of their data they began to believe that oceanic rocks must be created and destroyed in such a way that only young rocks remain today.

In 1962, Harry Hess advanced a theory called **seafloor spreading.** Hess's bold idea argued that the deep rift zones in the ocean basin were sites where the mantle had welled up near the surface and new crust was created. (This does not require that the earth expand through time if, as we now believe, oceanic crust is consumed elsewhere at a rate equal to that at which it is produced at mid-ocean ridges.) In 1965, a Canadian geophysicist, J. Tuzo Wilson, and an American geophysicist, W. Jason Morgan, independently offered the core of what geologists have embraced as the theory of plate tectonics. Other contributors to the ideas used in the new theory include Robert Dietz, Brent Dalrymple, Drummon Mathews, Lawrence Morley, and Fred Vine. This is only a partial list of contributors, for plate tectonics is an example of a framework that thrived on different types of data gathered by many different people and on similar ideas germinating in many minds at about the same time.

Taken together, it became quickly clear that continents that had seemed stable, capable of only limited and vertical movement, had indeed moved great distances across the globe (Figure 7-3). In retrospect, it might be said that Wegener's theory should have been given more serious consideration earlier. But like the idea of the organic origin of fossils, Wegener's original theory contained many daunting facets.

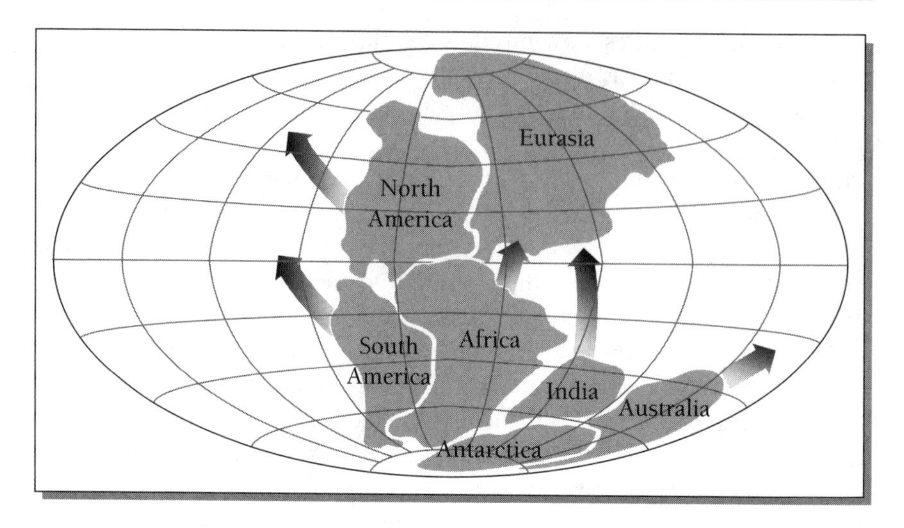

FIGURE 7-3: Pangaea. The arrows indicate the direction of movement as Pangaea broke apart in the Mesozoic Era.

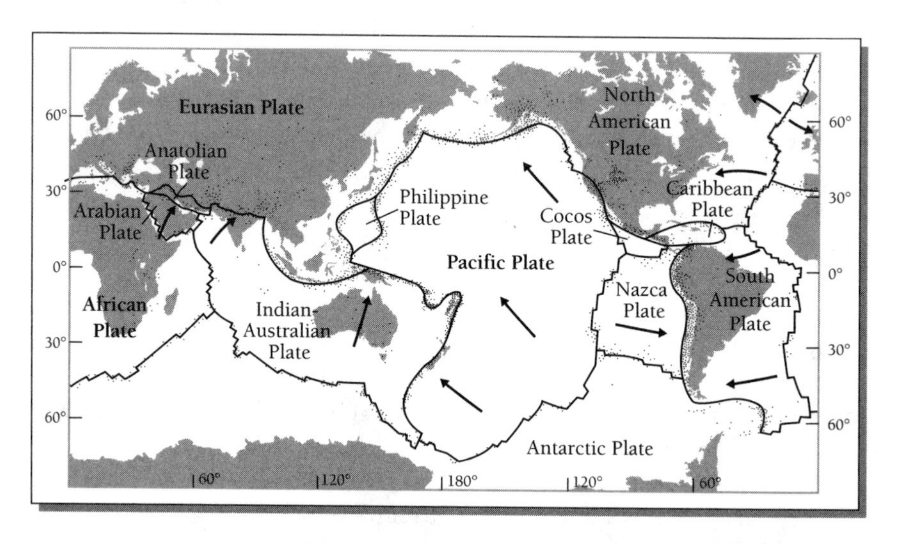

FIGURE 7-4: Earthquake epicenters. (Adapted with permission of Frank Press and Raymond Siever, *Understanding Earth* [New York: Freeman, 1995], fig. 18.15.)

The tectonic framework for interpreting the deformation of the earth's crust gained more and more supporters year by year because other lines of evidence—principally the remnant magnetism in rocks—fit so well with the theory. In addition, data on the distribution of earthquakes around the world made clear that seismic activity was located along bands that previous theories could not explain but that tectonic theory could (Figure 7-4). This kind of explanatory power made scientists around the world take notice and accept the new framework for geologic analysis.

But resistance to Wegener's ideas had been, at the time he proposed his view, quite reasonable. The magnitude of the forces required for physically moving the continents (no matter how gradual the movement) was a great drawback to any Wegenerian theory. Some mechanism for the forces needed to be identified. Geological ignorance of the ocean floor was an enormous gap. The correlative fossil evidence that Wegener and his followers often referred to was sufficiently complex that they made some false statements and confused nonspecialists.

The resistance to the radical and nonintuitive ideas of continental drift was to be expected. But given sufficient time and generations of geological researchers working largely without reference to broader themes, more and more evidence came to light in support of Wegener's basic idea. The data came from many different parts of the scientific spectrum. Drift theory, once it was more firmly rooted in plate tectonics, was thus rapidly accepted.

THE THEORY TODAY

According to a simplified view of tectonic theory, the uppermost portion of the earth is divided into **continental crust** and **ocean crust,** either or both of which ride on plates. These plates slowly move with respect to each other, with the movement creating three types of plate boundaries:

1. Divergent boundaries. In the oceans, at the ridges beneath the sea, molten material moves upward, thus creating the volcanism of the midocean ridges and such high points as Iceland. Plates are formed as the melt cools, with the plates moving apart from each other, away from the ridge (Figure 7-5).

2. Convergent boundaries. Since plates are created at ridges and move away from each other, there must be places where plates collide (Figure 7-5). These convergent zones are of two types: Either they form subduction zones of down-traveling oceanic crust, with associated volcanic mountain chains, or they form enormous mountains when two bodies of continental crust are involved in the collision.

3. Transform boundaries. In some places, the plates move past each other. This kind of boundary is dominant in California today and is responsible for the relatively shallow earthquakes familiar to residents of that state.

The movement of the plates has now been measured, and it is about as rapid as the growth of your fingernails (a very few inches per year). In Figure 7-6 the plates are sketched, and the magnitude of their relative movement is shown. Notice that the principal ideas behind modern tectonic theory fit nicely with

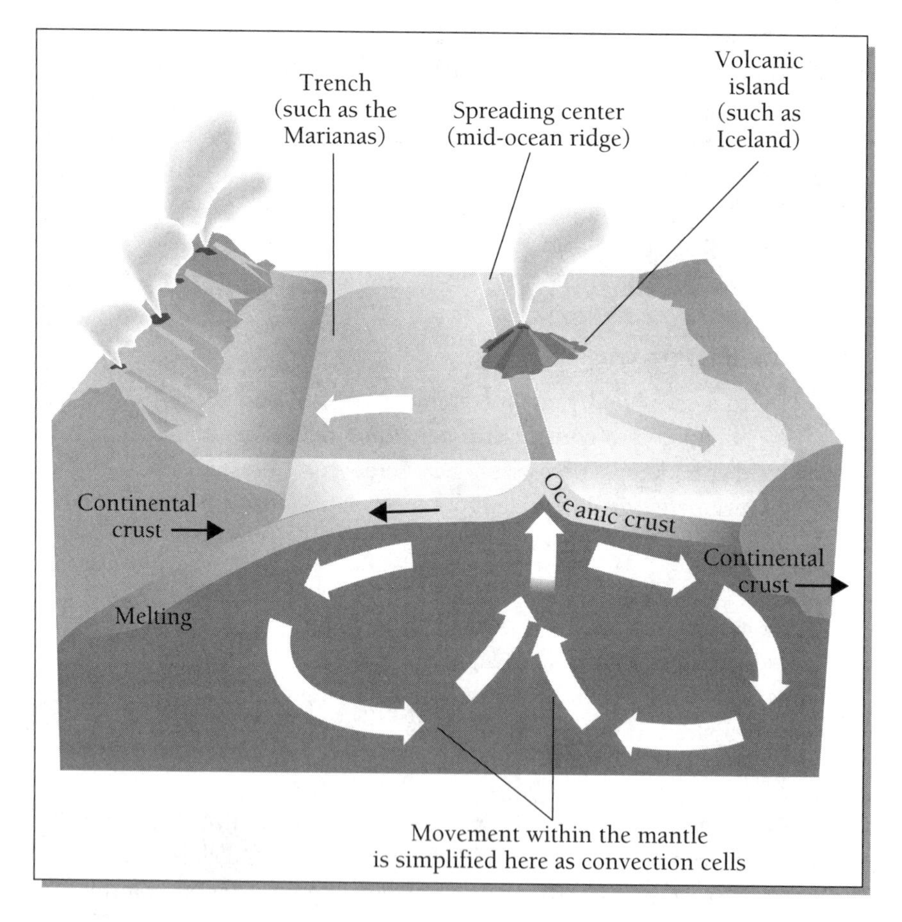

FIGURE 7-5: Deep movements beneath the plates are complex, but some may look like these.

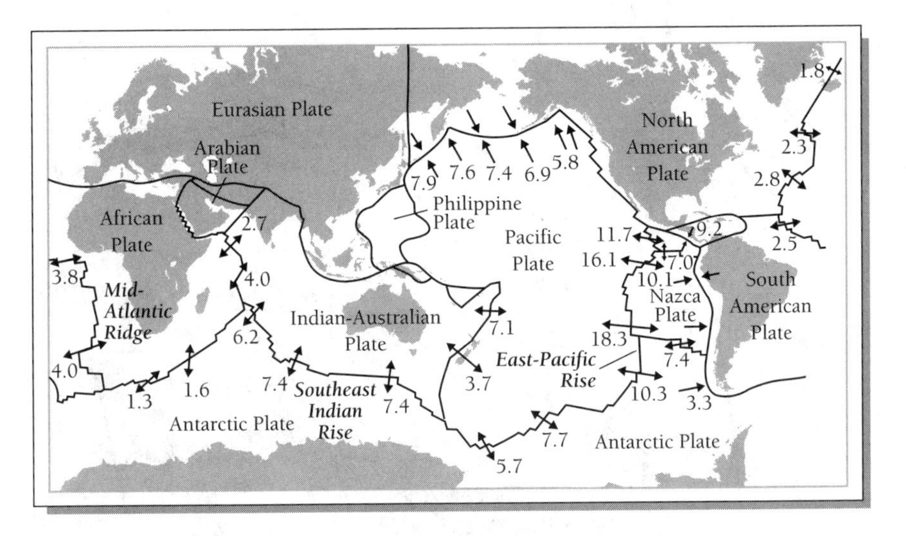

FIGURE 7-6: Average rates of continental movement in cm/yr. (Adapted with permission from Sheldon Judson and Steven M. Richardson, *Earth: An Introduction to Geologic Change* [Englewood Cliffs, NJ: Prentice-Hall, 1995], fig. 11.31.)

gradualism (even if tiny "jerks" in the plates' motions lead to the earthquakes shown earlier).

What accounts for the force behind the movement of the earth's massive plates? In the 1960s, geologists assumed that *convection* (the circulation of hot fluids upward and cooler portions downward) of the inner parts of the earth must account for the necessary forces. But geologists now believe that the interior portions of the earth are too *viscous* (stiff) to convect. Other mechanisms have been proposed, but none is fully satisfactory. Consider, for example, the spreading ridges that ring Africa. They seem to indicate an upwelling of material under the ocean floors, which would lead us to think Africa is experiencing a force pushing it inward in all (or almost all) directions. But Africa has a major *rift* (an early stage of a spreading center) in its northern sector, which seems to indicate upwelling beneath Africa as well as around it. More surprisingly, geologists now think hot upwelling is occurring under northern South America. This zone is now thought to have moved with the continent, staying just under it, for more than 100 million years. In short, the movements of inner earth are not understood, and we therefore cannot be fully sure what causes the plates to move!

GEOLOGICAL DOCTRINE: WHAT TO TEACH STUDENTS

The wavering of American geologists in first rejecting and then accepting the overarching theory of continental drift and plate tectonics was reflected in the revised freshman-level textbook by Leet and Judson, for years one of the best-selling of all such books. In the first edition, published in 1954, Leet denounced any notion of drift:

> The most spectacular and least plausible theory [to explain mountain building] which included among its panacean claims an explanation of certain mountain structures was that of A. Wegener relative to the

> [drift] of continents. . . . [T]he fit [of the outlines of continents that Wegener proposed] was more apparent than real and in detail was no fit at all. . . . [N]o forces even approximately adequate to produce [continental] drift have ever been suggested. . . . [I]t is claimed that the [mountain chains in western] North and South America were crumpled along the front edges of the drifting continents. As a matter of fact [these mountains] constitute positive proof against drift. If rocks of the ocean floor are weak enough to permit drift, there could be no folding of the advancing continents. If rocks of the ocean floor are stronger than the continents, there could be no drifting.[3]

By the third edition (1965) of the same text, however, the junior author had added a full chapter's discussion of continental drift and gave a balanced treatment of the disputes surrounding the theory. All subsequent editions of the textbook preach without reservation the gospel of plate tectonics. Since textbooks often tend to lag behind research, the clear change in Leet and Judson's third edition in 1965 was surprisingly timely.

It is important to remember, however, that geology is a largely descriptive, historically bound science and that its main strengths were never derived from widely encompassing theories about the planet as a whole. Even a quick reading of textbooks written before the 1960s shows that students were presented with a wealth of organized information about all branches of geology. The causes of mountain building were not understood (as the first edition of Leet and Judson, for example, clearly states). But a great deal about the earth's place in the solar system, about the formation of rocks and minerals, about the enormous length of geologic time, and about glaciation, river systems, and earthquakes was known. All this was taught to willing students without any difficulty. It was good and useful knowledge, even without any convincing theory of why mountains rose from the plains.

It is interesting that freshmen geology books today tend to skip over the problems of exactly what drives plate movements and how the inner earth relates to these plates. Textbook authors and instructors are generally reluctant to acknowledge the significant gaps in our understanding of these issues. In my view, this reluctance is a great loss for all concerned.

[3]L. Don Leet and Sheldon Judson, *Physical Geology,* 1st ed. (New York: Prentice-Hall, 1954), p. 292.

THE POWER OF A THEORY

Much of geology functions as a descriptive and observational science rather than an experimental one. Although its progress is uneven, geologic research has given us a very powerful and elegantly simple theory for the movement and deformation of the earth's outer portions. The theory of plate tectonics has helped explain linear belts of metamorphism, the migrations of sea turtles to distant islands, the young age of the oceans relative to that of the continents, and fossils that reveal mild climates in currently polar regions. Nevertheless, the research done before plate tectonic theory is still useful. Indeed, the concept of geosynclines is still accepted, though in modified form. It has survived because it is a useful way of organizing descriptive information about belts of sedimentary and metamorphic rocks.

Plate tectonics was rapidly accepted in the late 1960s because the data in its favor came from so many different types of research and because, in the end, tectonics told the best story about the formation of the earth's crust. The argument about continental drift was always, at its core, a historical one. In the direct method of physical science, historical events cannot be proved to have occurred in a particular manner. Geologists can gather only circumstantial evidence in support of a view. Such data are often noisy, or like the fossil evidence used by Wegener, they can support conflicting views. Arguments among specialists often weigh down the broader discussion. The best that geologists can hope for when they address a new and sweeping theory is that they will find a lot of circumstantial evidence from many different subfields of the discipline, and that this evidence will largely accord with the new theory. There can always be doubt—indeed quite reasonable doubt—on some points. Unlike the timeless physical sciences, which can invite observers to watch repeated experiments, a historical science like geology must settle for telling a powerful tale that sits well with most of the available, but always disparate, evidence.

SUMMARY AND CONCLUSIONS

Alfred Wegener's idea of mobile continents did not meet with universal acclaim in his day, even though the theory could be considered gradualistic. Geologists had almost no data regarding the seafloor, and Wegener, for his part, could not propose a mechanism for moving solid continents. If the earth were expanding significantly, some argued, continents could seem to move apart as the ocean floor grew between them. This creative idea, however, was set aside when more data became available.

To formulate a large-scale theory like tectonics, researchers had to summon observations from many different subfields of geology, ranging from the distribution of fossils in South America and Africa to the magnetic signatures of rocks in the middle of oceans. No one person can master all the details of specialized research, so it is easy to see why big theories invite either big mistakes or big successes. Naturally, debate between people holding opposing scientific views can be intense and, sometimes, intensely personal, but the result of geological debate has been the rapid advance of knowledge about our planet.

Today geologists accept the theory of plate tectonics. The theory argues that something deep within the earth provides the force that moves large segments, called plates, of the earth's outer layer. Some of these plates are wholly oceanic, whereas others carry both continental and oceanic crust. Many different types of data are in accord with tectonic theory.

QUESTIONS FOR DEBATE AND REVIEW

1. Say you are an American geologist in 1932, and because of the Great Depression you have lost your job. To pass the time, you gather all the ideas opposing Wegener's theory of continental drift. Describe the various issues.
2. Say you are a young geologist living and working in Iceland in 1972. What evidence might weigh most heavily with you in favor of plate tectonic theory?
3. Was Alfred Wegener intellectually negligent when he proposed continental drift without any mechanism for continental movement? How was he like J Harlen Bretz in this regard?
4. In regard to Question 3, was the idea of an expanding earth, as proposed in the 1940s and 1950s, actually so absurd? Does it look different to us today because the idea did not prove fruitful?

CHAPTER 8

A Huge Gap in Gradualism: The Genesis of Life on Earth

The earth's most unusual and intriguing feature may be that its surface is covered with a multitude of plants and animals thriving in the oceans and on land. Humans have probably always been interested in life's origins; supernatural explanations for the birth of life are found in cultures around the world. In the modern era, scientists have addressed the same set of questions that mythology earlier tried to answer. Scientists, of course, have used the tools of reason and experiment to forge an understanding of life's origins. Although scientists have made much progress, they are still far from being able to explain how complex organic molecules like DNA and RNA could have arisen in the earth's early history.

The Traditional Approach

Although it may seem strange today, for most of human history people thought that new life was generated from inorganic matter on-going active processes around us. Aristotle and his intellectual descendants in the Middle Ages accepted without question the idea that simple creatures were created directly from soil and rock by the force known as *latent plastic virtue*. After all, worms, insects, algae, and other simple life-forms appeared to emerge each spring from the dark, cold muds of lake bottoms. Naturalists believed that such life was spontaneously generated by the basic creative force, plastic virtue, which we encountered in the figured stones debate of Chapter 3. Indeed, the spontaneous generation of simple life-forms was accepted even well after the beginnings of the modern age. It was an idea in accord with the Hutton–Lyell framework of uniformitarian gradualism that emphasized continuing processes. Scientists and laypersons alike thought that "germs" could come into existence anytime, at least in welcoming environments.

The famous French scientist Louis Pasteur (1822–1895) overturned the notion behind the spontaneous generation of life. His was a long battle, for the idea of completely separating life from the inorganic realm was fundamentally new. Pasteur argued that only living creatures could beget living creatures, no matter how simple. In a series of experiments he showed that no life-forms, not even bacteria, spontaneously arose in sterile conditions. Even a broth rich with nutrients for growth showed no tendency to spawn life if it had nothing living in it. For the first time, scientists concluded that the creation of all life-forms, from the simple to the complex, had occurred sometime in the past. With regard to the generation of life itself, it appears that we do not inhabit a world shaped by continuing processes.

Doubting Initial Assumptions

It always is good intellectual exercise to doubt one of the assumptions of modern scientific work, so let's question what has been taken for granted since Pasteur's time. Is there, after all, a possibility of spontaneous generation of life in the present? After all, if life did arise spontaneously on the early earth, why couldn't it do so now, under similar conditions? Researchers initially wondered about this question in connection with new and exotic ecologies discovered on the ocean floor. Some scientists speculated that perhaps the hydrothermal (hot-water) vents at mid-ocean ridges may have provided an environment that could generate life. Like the early earth, the vents are hot and supply reduced (nonoxygenated) sulfur, nitrogen, and carbon to the water around them. From what scientists can tell, the multicellular animals living in communities at the vents exist nowhere else on the planet. Why not imagine that Pasteur was a little hasty? After all, his experiments did not include this sort of environment, the kind in which modern scientists think life did arise of its own accord.

The animals living at marine hydrothermal vents are indeed not found elseswhere. But investigation has shown that the genetic code of life at the vents is not fundamentally different from our own. DNA and RNA are the backbone of heredity in such communities, and since DNA and RNA are highly complex molecules, scientists doubt that they would have been separately produced in two different instances and yet be exactly the same. It is vastly more likely that the vent animals and those more familiar to us surface dwellers all have common ancestors in the Archean eon.

> The genesis of all life, it appears, was in the past, just as Pasteur argued in the last century. Exactly why this should be so, however, certainly remains open to question, and the implications of the assumption that life could be created only once on this planet have not yet been explored fully. That's quite a gap in our thinking!

The Birth of Life

HOW DID IT BEGIN?

Geologists are conditioned by their natures to think about the origin of life in terms of efficient causality. What physical conditions, they ask, would have been favorable to the earth's earliest life-forms? What is the earliest evidence of life in the fossil record? More fundamentally, biochemists speculate about how organic molecules may have arisen from inorganic components. Once they did arise, of course, there was much more to accomplish in generating the highly complex structures of the living cell.

THE REQUIREMENTS FOR LIFE

Biologists tell us that in order for a thing to be alive, it must satisfy the following requirements:

1. Cellular structure. All life on the earth, from the bacterium to the elephant, is composed of cells. We do not know why this is the case, but cells are ubiquitous in the biosphere.
2. The metabolic assimilation of energy. All living cells consume energy, which powers the chemical reactions inside them.
3. Reproduction. All organisms, from the simplest to the most complex, reproduce. (This normally offsets the death rate of members

of a given species; for unknown reasons, both death and reproduction are woven into the fabric of all life on earth).

4. Heredity. All life-forms on earth pass on their own characteristics to the next generation, using genetic information encoded in and transmitted through the nucleic acids DNA and RNA. Again, we do not know why this is the case, but life on earth uses these two molecules to transmit characteristics to offspring.

The first question a scientist investigating earth history might ask is, When did organic molecules achieve the organization to meet these four criteria? The earliest life that geologists have discovered in the rock record dates back to the Archean eon, the oldest time from which rocks are preserved. It is not surprising, perhaps, that the earliest fossil life is simple. Prokaryotes, better known as *bacteria*, are the oldest fossil life-form identified on earth. Their simplicity is indicated by their lack of a cell nucleus and the fact that the DNA of bacteria is not organized into chromosomes. There are several major types of prokaryotes or bacteria:

1. The most primitive are the anaerobic, methane-producing bacteria. Geologists believe these have been on earth at least as far back as the early Archean eon.

2. More complex are the bacteria using photosynthetic reactions to generate energy, similar to the actions of plants. They include the blue-green algae, more properly called *cyanobacteria*. We find them in the fossil record as stromatolites well after the appearance of the methane-producing bacteria.

3. Last are the younger and most complex bacteria, including the forms perhaps most familiar to us, for example, the bacteria living symbiotically in our intestines and the bacteria which cause tuberculosis and gangrene.

THE FOSSIL RECORD

Geologists find prokaryotes preserved in some of the oldest sedimentary rocks on the planet. But even the simplest bacteria are extraordinarily complex compared with inorganic structures. How did the complex, organic building blocks for bacteria cells come into existence?

Remember that the origin of life in geologic history appears to be a truly unique event. Scientists think that life began at one point, or at least during one interval, in the planet's past. Consider carefully what this short statement means: Life did not exist on earth before a particular time, and life has not been recreated in any ongoing processes that we might be able to observe or manipulate. By definition, unique events are not open to experimental investigation in the way that continuing processes can be.

Sudden and unique events, one might say, are the extreme and limiting case of catastrophic change: Things were different before the Event; then the Event occurred; and now we cannot hope to see the Event in action ever again.

To put this point in another way, we could say that the origin of life is embedded in history, and the past is never open to direct experimental investigation. In this regard, remember that no matter what scientists do in the laboratory to mimic the conditions of the early earth, they cannot prove that a given organic molecule or reaction path was the one important in a particular time and place 4 billion years ago. They may be able to say a great deal about which paths are likely to have been more important than others, but that is as far as they can go. Given these considerations, it may not be surprising that the apparently unique event of the origin of prokaryotes has led scientists to consider several highly different theories about the earliest history of life.

What were the physical conditions that the earth's first life would have encountered and from which it may have sprung? The planet, geologists are sure, was born in a molten state. In the early Archean eon, the world had cooled to the point that solid rock and liquid water could exist. The rock record indicates that the earth's early atmosphere lacked free oxygen. Like Venus today, the early earth was probably cloaked in an atmosphere composed principally of carbon dioxide. This absence of oxygen meant, among other things, that the earth had no ozone layer (ozone is composed of oxygen in the form O_3). Without an ozone layer, scientists think that a great deal of ultraviolet (UV) light from the sun must have penetrated the atmosphere and struck the surface of the earth. Because UV light destroys living cells, scientists therefore reason that the early life-forms lived in water, which helped block the UV light. But intense UV radiation throughout the Archean eon may have been important to the origin and early shaping of life.

THE EARLY EARTH

Geologists think that there was probably ammonia, hydrogen sulfide, and methane in the early atmosphere, spewed into the atmosphere from volcanic eruptions just as they are today on a lesser scale. These gases contain the chemically reduced forms of nitrogen, sulfur, and carbon, all of which are central to life as we know it. In the famous chemical experiments of Stanley Miller and Harold Urey, boiling water contained in a vessel with these gases was subjected to an electrical current, meant to simulate the effects of lightening. Within one week's time, the boiling water was capped by a fluid layer of organic molecules. Although the chemicals were "organic," they were not, of course, "alive." But their presence was a great discovery and for many years spurred research into life's origins.

The methane gas in the Miller–Urey experiment had broken down into formaldehyde and hydrogen cyanide, both of which are highly reactive molecules. Apparently these agents produced complex organic molecules, including four amino acids. The Miller–Urey experiments inspired other attempts to construct organic molecules from the ingredients of the primitive oceans and atmosphere. Other scientists have created a ring-shaped molecule called *adenine*, one of the bases of DNA and RNA. This is an exciting result, but beyond this point, experimental work has not progressed far.

THE INTRODUCTION OF DNA

The amino acids produced in the Miller–Urey type of experiments sometimes react to form proteinoids. Proteinoids have been experimentally shown to form **coacervates**, or tiny spheres like soap bubbles. These spheres have special properties. They have the same shape as cells, and *cells,* remember, are the first of the requirements that biologists stipulate for life. The coacervates can also contain amino acids that help decompose glucose. This is an energy-releasing process, satisfying the second requirement for life, that of *metabolic energy consumption*. The coacervate spheres sometimes expand and then subdivide into two new spheres, satisfying the third requirement of life, that of *reproduction*. But the coacervate spheres are not considered to be living creatures because there is no heritable component in their reproduction. That is, they lack DNA, RNA, or any similar molecule that could pass on characteristics to a new generation of coacervates. Still, some scientists think that coacervates may have been an important step toward truly living cells. Some researchers posit that RNA and DNA somehow came into existence in the proteinoid spheres, and then life as we know it began.

This, at least, is one view. But because scientists are grappling with a question without clear empirical constraints, they have taken some highly different approaches. Some scientists take an aggressively inorganic approach to the problem and nominate clay particles for the role of primitive self-duplication. Some clay surfaces do, in fact, have self-replicating processes. In time, according to this hypothesis, organic molecules used clay structures as a kind of template for organic self-duplication. Then for unknown reasons, the organics were diverted down a wholly carbon-based path toward prokaryotic life, as seen in the fossil record.

AN ASIDE

Viruses are very primitive life-forms, composed of a molecule of RNA or DNA inside a protein membrane. Some of them can be dried, crystallized to a salt, and then "revived" with water at a later time. They are, perhaps, the borderline between the inorganic and the organic realms. But viruses, scientists believe, could not have been the first life on earth, because they are parasites of other cells, reproducing only within a living host. But could primitive protoviruses have reproduced within coacervates? Unfortunately, geologists have nothing in the fossil record before the appearance of full bacteria cells to dampen such speculation; many different possibilities come to mind, and it is difficult to know how scientists might choose among them.

INFORMATION THEORY

A different way of thinking about the origin of life is to consider **information theory**. This approach concentrates on the structures of nonrandom relationships, exactly the kind of relationships inherent in life. According to this point of view, the genetic code of DNA is an extremely long arrangement of molecules in a highly particular order. Each strand of DNA represents a great deal of information, the opposite of random relationships or "noise" (Figure 8-1).

It would seem that scientists can consider two possibilities regarding the origin of biological information:

1. Simple chance led to the earliest life-forms after Miller–Urey type reactions had generated organic molecules in the earth's ancient oceans. The probability of such an event seems tiny, but as the argument goes since life exists, it must have occurred. (Or to put it crudely, once you know you won the lottery, the odds against you don't have much meaning for your case.)

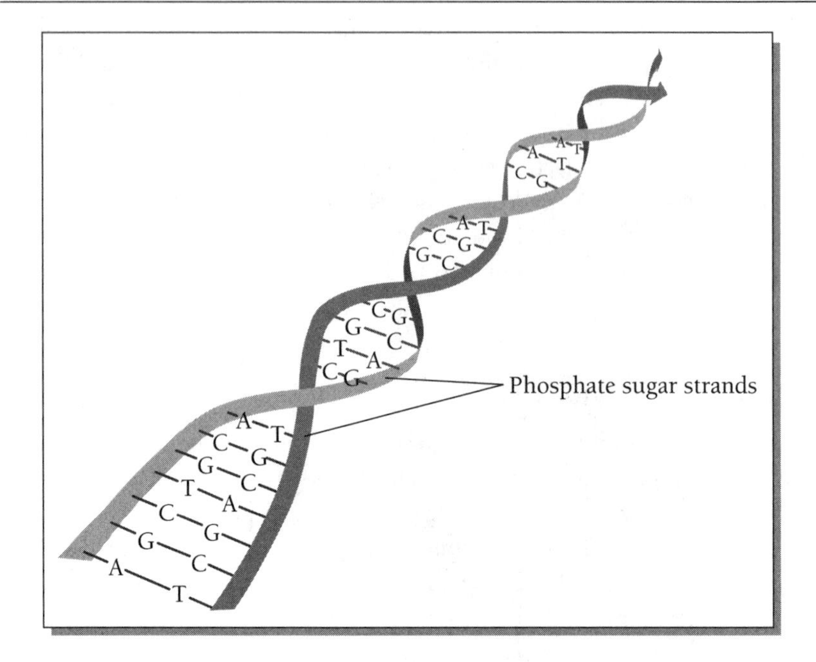

FIGURE 8-1: The double helix of DNA. Genetic information is stored in DNA molecules simply by the order of Thymine (T), adenine (A), cytosine (C), and guanine (G). The molecule is very long, so the amount of information that can be stored is large—large enough that your DNA is different from that of all your siblings (excluding identical twins). The DNA molecule is so unlikely to have arisen solely by chance that many scientists assume the existence of some kind of prebiological evolution (and many theologians assume God's guiding hand).

2. There must have been some sort of "evolutionary" process that shaped the nature and abundance of large and complex organic molecules before life began. These organic molecules may have reached the stage of self-replication before true life began, with natural selection favoring some over others.

The problem with the first idea, that of the random construction of DNA and RNA, is that the chance of such an occurrence remains remote, even though life evidently "won the lottery." In any event, it is poor scientific methodology to use the idea of chance to explain puzzling natural events. Indeed, one of the points of scientific research and the concentration on efficient causality is to try to minimize the role of chance in scientific explanations of natural phenomena. These methodological considerations drive scientists toward the second idea, that the extraordinarily rich information systems in cells were shaped by a natural weeding-out process. The mechanism of selection must have included some sort of self-replication and nonbiological evolution before true life began. Although this is a very promising line of thought, research at present into such ideas is in its earliest stages.

SCIENCE FICTION IN REAL LIFE

At one end of the continuum of debate is the idea of **extraterrestrial** influences, an idea that has a long history but is highly speculative. Theories about life's origins that appeal to alien life-forms range from simple fiction to intellectually grounded hypotheses. Francis Crick, who was awarded a Nobel Prize for his part in discovering the structure of the DNA molecule, is the author of a book arguing that life was deliberately sent to the earth by advanced organisms elsewhere in the galaxy. Crick calls his idea *panspermia* (from ancient Greek words meaning "the seeds of living things are everywhere in the cosmos").

Crick notes that the earth was born many billions of years after the universe originated, and he reasons that sufficient time was available for life to originate (somehow) elsewhere and become highly complex. Aliens may have sent primitive life-forms to many planets possessing the basic conditions for life, hoping to seed them. Bacteria, Crick thinks, are one likely candidate for the interstellar seeds, since they are quite hardy in extreme conditions and might have survived the journey in space. Crick speculated that aliens may have wanted to plant life on earth so that the atmosphere would become oxygenated for their later use.

Until and unless the presence of extraterrestrial life is confirmed, Crick's theory is wholly speculative. At least in part, his motivation in offering the panspermia hypothesis was the intellectual challenge of building a case for the idea in the face of a paucity of data about a pivotal, unique event. Not many scientists have given Crick's book serious consideration. But in fairness to his hypothesis, some of the well-known materials alien to the earth, like meteorites, are not the simple inorganic silicates and metals we might expect in space.

One type of meteorite, the carbonaceous chondrite, contains carbon in complex, organic molecules. These compounds even include amino acids. Crick goes further than other scientists who study meteorites when he maintains that microscopic shapes in some samples represent primitive life-forms that "live" on the metals of the meteorites. Most scientists interpret the shapes as inorganic in origin or as biological contamination from our own planet after the meteorite fell to the earth's surface.

Although a few scientists are taking seriously the idea of life elsewhere, most point out that any extraterrestrial origin for life only pushes back the fundamental questions. Why did these hypothetical life-forms come into existence? Obviously, scientists cannot begin to address a question concerning life-forms for which they have no evidence.

Indeed, it could be said that Crick's hypothesis is not unlike claims that a supernatural being created life by means of divine will. Neither idea points scientists toward much work they could do to verify the claim, and neither gives science any explanatory power regarding life's origins. Perhaps aliens elsewhere in the galaxy did launch life-bearing vessels of "seed" in our direction, or perhaps the Judeo-Christian God did miraculously create the first germ of life. But how can scientists investigate these ideas empirically? The spirit of scientific research requires the search for observations and experiments to verify or falsify ideas.

At first glance there does not seem to be much functional difference between Crick's space aliens and the God of Genesis. But in any event, why shouldn't life have begun here? At least we do know that there is life on earth, which is considerably more than Crick knows about his space aliens on other planets in our galaxy. There is simply no reason, most scientists believe, that life could not arise of its own accord, right here on earth.

The View of Religion and Science on the Origin of Life

SCIENCE LOOKS WEAKER THAN USUAL

As we have seen, the earliest history of life remains unknown, and scientists do not yet understand the origin of life on our planet. So far, scientists have not offered any single theory for the generation of life that is convincing to many members of the scientific community.

Defenders of religious faiths often dwell on scientists' inadequacy on this point. It is not surprising to many theologians that scientists have made little headway in understanding the genesis of life. All Western religious traditions teach that life has a supernatural origin that is a direct reflection of the hand of God. It is understandable, perhaps, that theologians like to think that science will not be able to determine how such complex molecules as RNA and DNA were derived from the primordial oceans.

Scientists, however, point out that they discovered RNA and DNA only one generation ago. In a few more decades, a great deal more will be known about biochemistry, and it is quite possible then that an inorganic explanation of life's origins will be convincingly developed. Religious critics can hardly deny the surprising rapidity of research in organic chemistry. But they respond that although scientists may discover what they confidently hope for, they also may not. The point seems valid, and only time will tell what biochemical research will accomplish. In any case, the matter will never be proved, for even if scientists find a way of producing RNA and DNA from inorganic constituents, their laboratory mechanisms may have little or nothing to do with the way that events in the Archean eon actually did unfold. Historical science always brings with it this important caveat.

Final Causes Again

The history of life, theologians sometimes note, is a record of increasing physiological sophistication. Organisms have become more and more complex, both individually and in terms of their roles in social units. The cause of this trend is opaque to science, theologians sometimes argue. But of course, it is obvious enough from a religious perspective. God, one might assert, is directing the course of our biological history toward his chosen final cause. God's contribution to the evolutionary mix can never be detected by science in any certain way. Science may find a few conditions in the natural world to be odd, fortuitous, or highly improbable, but nothing that science can discover will directly reveal God's intentions.

In secular terms, both life and the universe may be evolving toward something. As the physicist Paul Davies wrote, "The universe has never ceased to be creative. . . . [R]esearch in areas as diverse as fluid turbulence, crystal growth, and neural networks, is revealing the extraordinary propensity for physical systems to generate new states of order spontaneously. It is clear that there exists self-organizing processes in every branch of science."[1]

This perspective puts life's origins and evolution into a broader framework, and it can lead to the idea that the goal of our world is increasing self-organization or increasing self-consciousness.

Summary and Conclusions

Scientific work can address only certain types of questions. Aristotle and early scientists urged their students to formulate hypotheses in such a way that evidence could be gathered for or against them. The interstellar explanation for life's origins as put forward by Francis Crick may well be outside the

[1]Paul Davies, *The Cosmic Blueprint* (London: Heinemann Press, 1987), p. 1.

boundaries of science, because there is little that scientists can do to test his panspermia hypothesis. Researchers are on firmer ground with Miller–Urey types of experiments. Clearly, under the right conditions, organic molecules, including amino acids, can be generated from inorganic chemicals. Furthermore, the conditions in question are in broad accord with what geologists believe the early earth was like.

But scientists still have a long road to travel in understanding how RNA and DNA might have arisen. The possibility of the natural selection of organic molecules in the earliest ocean is intriguing, but it seems to be beyond verification. Perhaps scientists are at the edge of what science can usefully consider. Theologians generally agree with that sentiment and believe that science cannot, by its nature, fully address the origin of life. Religious writers usually appeal to the concept of a final cause to express their views on the genesis of life.

But once cellular life began, geologists can clearly trace the development of simple life-forms during the Archean and Proterozoic eons. Eventually, near the end of the Proterozoic, multicellular life appears in the fossil record. The faunas of the Paleozoic era are complex and diverse, more similar to what we know today than anything that had existed in the first 3 billion years of the earth's history.

QUESTIONS FOR DEBATE AND REVIEW

1. Can Francis Crick's idea be tested in any way? What could we do to give it some empirical support? What kind of evidence concerning the origin of life would you find convincing? (Remember that work in historical science cannot be proved, that we cannot return to the past and repeat the "experiment." Crick's theory must thus rely on something less direct—and less incontrovertible—than the testimony of an unimpeachable eyewitness.)
2. Does pointing to simple chance for the origin of life (that is, complex organic molecules just happened to bump into one another in a way that led to life) seem to you to be an intellectually responsible belief? Explain.
3. This text presents the "definition" of life given in biology textbooks. Some philosophers would object to this, saying that they are the people who should wrestle with a definition of life. If you watch science fiction movies, you are familiar with some of the difficulties in answering the question, Is this or that weird thing alive? Give some examples from the

popular culture of entities that are alive but do not fit this text's definition of life. What are the important properties that define "life" in light of these examples?

4. Explain and defend a theory of the origin of life on earth.

CHAPTER

9

The Fossil Record: A Biased but Rich Record of the History of Life

Unlike the other planets in our solar system, the earth is teaming with life. Wherever and however it may have arisen, life has been successful on this planet. For more than 4 billion years, different organisms have flourished, first in the oceans and then on land. The history of life is, in itself, an absorbing story. A record of this history is given to geologists through the fossils contained in sedimentary rocks.

Are Fossils Good Witnesses?

Unfortunately, the fossil record does not reveal everything that happens on the earth, because when an individual plant or animal dies, it is only rarely preserved as a fossil. Instead, the organic materials bound up in the bodies of plants and animals usually are recycled (eaten by predators and scavengers or decomposed by microorganisms) shortly after they die. This is especially true for species living on land, where there are few mechanisms for physically covering organisms after death. For example, the environments where lions, tigers, and bears live have little sedimentation. Indeed, most of the surface area of terrestrial environments is suffering from erosion, not deposition. Lion, tiger, and bear carcasses are therefore usually recycled quickly. Only if a tiger, say, dies at the edge of a lake or stream into which it then falls, will its bones be buried in the kind of environment where fossilization is possible. A few saber-toothed tigers of the Pleistocene had the courtesy to die in tar pits, an excellent environment for preservation, but in general, tigers are poorly represented in the fossil record, by virtue of their land-based habitat. The same, of course, is true for the species that often most interests us, *Homo sapiens sapiens* and its immediate predecessors.

The history of life is most reliably recorded in the **marine sedimentary record.** The sedimentary rocks formed in ocean basins are our best source for information because they are formed in geographically large environments of fairly continuous deposition. But even in the oceans, it is only those species with **hard parts** (shells or bones) that are likely to be preserved. Thus geologists have a much better record of the evolution and extinctions of shellfish than they do of jellyfish. Nevertheless, geologists do not have anything like a complete, annual record of the history of even marine species with hard parts. In short, the database for the history of life is far from ideal. As we shall see in later chapters, the imperfections of the fossil record must be borne in mind when considering the theories for both evolution and extinction in the history of life on earth.

The History of Life

One of geology's great triumphs is the description of the fossil record of life through its 4 billion-year history (Figure 9-1). We summarize this subject as follows (see Appendix B for the biological terminology and Appendix C for the terminology of geologic time).

Life in the Hadean Eon (4.6 billion to about 4 billion years ago)

The Hadean eon, named for Hades, the Greek god of the underworld of ancient mythology, corresponds to the earliest period of earth history. The planet was largely molten during this time, and the few igneous rocks that formed were remelted. We have no direct record of conditions on the earth in the Hadean period, but we assume they led gradually into the next eon, a time for which we do have a historical record, preserved in igneous, sedimentary, and metamorphic rocks.

LIFE IN THE ARCHEAN EON (4 TO 2.5 BILLION YEARS AGO)

Almost one third of the earth's history occurred in the Archean eon. Archean life was very simple. Bacteria and **blue-green algae,** now known as *cyanobacteria*, were present throughout the eon. These primitive cells, termed **prokaryotic,** had no nuclei. (Other, later life is composed of plant and animal cells that do contain nuclei and chromosomes; such cells are called **eukaryotic.**)

Graphite, which is composed of pure carbon, is found in some of the oldest sedimentary rocks of the Archean. Its presence may indicate that organisms lived in an environment of deposition, thereby concentrating carbon, which was later altered to graphite. Better evidence of life is given by stromatolites, which geologists find as fossils in Archean rocks. Stromatolites are a special, colonial algae still extant in some shallow seas. They grow in large mounds under the water as they precipitate calcium carbonate from seawater. Some Australian rocks dated at about 3.5 billion years show evidence of stromatolites. Although the record is scanty, it seems clear that very simple life-forms existed throughout the marine environment of the Archean.

LIFE IN THE PROTEROZOIC EON (2.5 BILLION TO 543 MILLION YEARS AGO)

The history of life changes noticeably in the Proterozoic eon, maybe because the earth's physical structure was changing. The planet had cooled to the point that a modern style of plate tectonics could begin, generating a felsic continental crust. The continental shelves around the cratons (stable continental crust) provided large and shallow seas for life to exploit. Such environments are characterized by the deposition of sediments in large quantities, and thus geologists have a fairly good record of life in such places.

As mentioned in the previous chapter, the atmosphere of the earliest earth contained no free oxygen. However, stromatolites throughout the Archean eon had been liberating oxygen, just as plants do today. A moderate level of free oxygen in the air was reached about 2 billion years ago in the Proterozoic eon, a change crucial to more advanced life-forms.

The history of life during the Proterozoic looks like this:

In the early Proterozoic, stromatolites become more abundant than ever, filling up the continental shelves. Free oxygen begins to appear in the atmosphere. Then, in the mid-Proterozoic, **eukaryotic cells and multicellular algae**

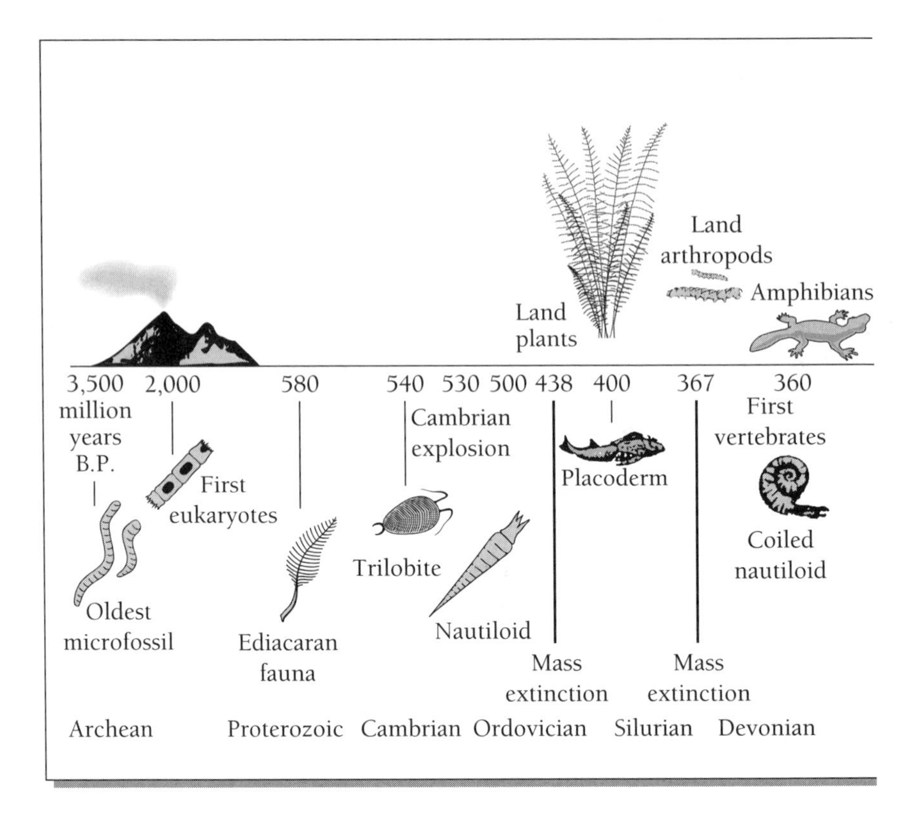

FIGURE 9-1: The emergence of life. (Adapted with permission from *Life in the Universe: Readings from Scientific American Magazine* [New York: Freeman, 1995], pp. 6–7.)

appear in the fossil record. Recently, geologists working in China have also found abundant, if strange, soft-bodied animals, dating to about 1.7 billion years ago. They are big (measured in inches and even feet) and reproduce by mixing the DNA of two animals, thus increasing the chances that some offspring will be substantially different from their parents. These animals are flat and segmented. Some look like pancakes, some like flat worms, some like disks. These organisms, geologists believe, were exchanging gases with the water directly through their bodies. They therefore had to be flat or shaped like a pencil so that each part of their bodies could "breathe." Nearly all later animals exchange gases through specialized organs (lungs and gills) and therefore need not be flat as a pancake or thin like a thread. The name

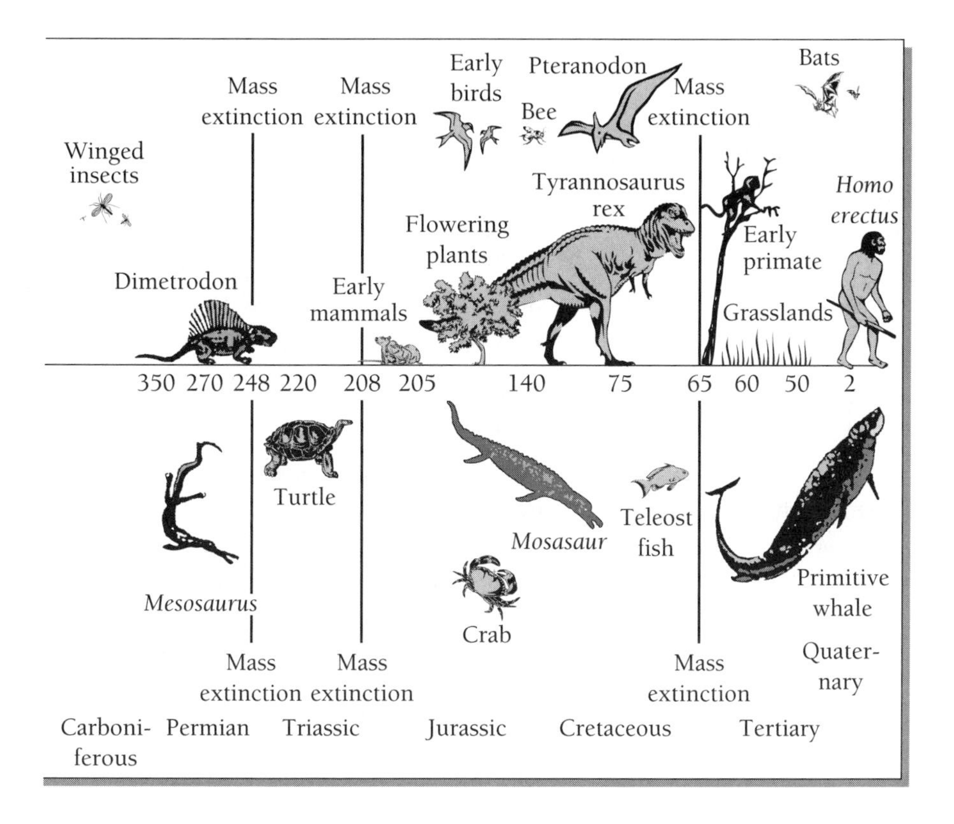

for the most developed faunas of this sort in the late Proterozoic is *Ediacara*, named for an area of Australia.

The Proterozoic eon records much more change in the history of life than the Archean does, and by the end of the Proterozoic (that is, the end of the Precambrian), the stage was set for even more rapid change. Complex animals reproduced by exchanging genetic material on chromosomes. Because of the marine plants, both single and multicelled, oxygen levels in the atmosphere reached levels near those of the present. The next eon was to see an explosion of diversity in life-forms. Geologists call it the **Phanerozoic eon**, from the Greek words meaning "visible life," to honor the dramatic change in the fossil record. Before the Cambrian period, fossils are small, indeed often microscopic, and few. But during and after the Cambrian, fossils are numerous and diverse. (See Appendix B for a brief survey of the vocabulary of complex life-forms.)

Life in the Phanerozoic Eon

LIFE IN THE PALEOZOIC ERA (543 MILLION TO 248 MILLION YEARS AGO)

The Paleozoic era begins with the Cambrian period. The rock record of the Cambrian is quite suddenly crawling with creatures.[1] For the first time, some animals were predators, feeding on herbivorous animals. We might say that Mother Nature's "arms race" had begun, with herbivores showing defensive mechanisms (like shells) and predators beginning to display body parts designed to kill and ingest their prey. Early Cambrian predators may have eliminated many of the earlier animals whose soft bodies and stationary lifestyles made them easy prey. Trilobites burst onto the scene in the Cambrian period. They were diverse and abundant and swam and crawled into many different depositional environments.

Furthermore, and luckily for geologists, many trilobite species were short lived. They therefore have become useful **index fossils** for the period, their presence denoting particular, fairly small, stretches of geologic time. Index fossils are important to geologists who have come to value any organism that makes it relatively easy to isolate a particular piece of geologic time. Fossils of this sort have the following characteristics: (1) They are easily preserved; that is, they have hard parts; (2) they were widespread and present in several different environments (for example, muddy shores, sandy shores, deeper water); and (3) they were short lived.

Brachiopods, double-valved suspension feeding animals, appear in the Cambrian, and mollusks are preserved in the rocks. A swimming, eel-like animal called a *conodont* arrived on the scene. Its teeth—also called conodonts—are the common fossil remains left in the rock record.[2] Primitive fish also appeared.

[1]The basic designs of some of the earliest complex animals with hard parts, that is, those best preserved in the fossil record, did not survive even into the late Cambrian. The Burgess Shale fauna are Cambrian animals with a symmetry and morphology much different from that of later life. The fauna became extinct for unknown reasons, and more "modern" patterns of life replaced them. Like the Ediacara fauna, the Burgess Shale fossils are a fascinating look at one of the paths down which life, for one reason or another, did not travel long. This subject is treated in detail by Harry B. Whittington, *The Burgess Shale* (New Haven, CT: Yale University Press, 1985). A more popular treatment is by Stephen Jay Gould, *Wonderful Life* (New York: Norton, 1989). A truly gorgeous photographic record of the Burgess Shale was produced by D. Briggs, D. Erwin, F. Collier, and C. Clark, *The Fossils of the Burgess Shale* (Washington, DC: Smithsonian Institution, 1994).

[2]The small, pyramidal fossils usually assumed to be the teeth of the conodont animal may instead be part of a structure inside the head or jaw. But they are usually termed *teeth* in introductory geology classes.

Geologists think that most Cambrian animals were herbivores that fed on algae. There apparently were just a few predators, such as the cephalopoda which had tentacles to hold small animals. Larger animals with hard shells (for example, the mollusks) may have had no effective predators.

There were several episodes of major extinction in the Cambrian, with the most severe coming at the end of the period. The ground was cleared, in a manner of speaking, to get the Ordovician period off to a different start. New animals in the Ordovician include the crinoids and bryozoans. The Ordovician period closed with a tremendous mass extinction that ended the "early Paleozoic" world. Although a few species of trilobites survived into the next part of the Paleozoic, they never again were dominant.

In the Silurian period, the bivalve and gastropod mollusks flourished, and corals grew throughout shallow marine waters. Fishes apparently moved from seawater into freshwater. In the Devonian period, new kinds of *nektonic* (swimming) animals appeared. Some had hard shells to protect the animals living inside, and so they are well preserved in the fossil record. In another important step, multicellular plants colonized the land for the first time. Shortly after that, it is thought, arthropods (insects) followed the plants to the land. Finally, sometime in the Devonian period, amphibians made the journey from the water to the land. Toward the end of the Devonian period, however, there was a mass extinction of marine life.

Next, the Carboniferous period—divided in this country into the Mississippian and the Pennsylvanian—was an age in which the eastern United States was dominated by large swamps. Large fernlike plants and trees flourished in shallow water, and their remains formed the coal beds we mine today. In addition to major changes in terrestrial plants, the Carboniferous saw the appearance of reptiles. For the first time, animals could live and reproduce away from water.

The Paleozoic era ended with another mass extinction, which may have been the greatest ever experienced in earth history. The trilobites, for example, abruptly disappeared. Both land and sea were cleared of many species.

LIFE IN THE MESOZOIC ERA (248 MILLION TO 65 MILLION YEARS AGO)

At the beginning of the Mesozoic era, it is believed that all the earth's continents were joined together in one giant landmass called **Pangaea**. In time, Pangaea broke up, and over millions of years the continents assumed the shapes familiar to us today. As these physical processes took place, life-forms also changed. Some reptile species appeared in the sea. New kinds of fishes appear in the fossil record, and mollusks flourished. The dinosaurs also made their debut and thrived in a wide variety of terrestrial environments.

Dinosaurs were mostly small and agile, moving quickly about on two legs. There were two main types: the "bird-hipped" dinosaurs, called *ornithischians*, whose bodies could be held directly above their legs. Those with hip bones more like those of lizards, called *saurischians*, had legs sprawled out from their bodies. All the ornithischians were herbivores, whereas the saurischians included both herbivores and carnivores.

The really massive dinosaurs that fascinate the public's imagination are called *sauropods*, and they moved about on all fours. Studies of the fossil record around the world are helping us understand these dinosaurs better. Geologists no longer think they were big, slow, and stupid. They sometimes traveled together in herds, and geologists have discovered nests of baby dinosaurs in upper Cretaceous rocks. The young lived together, many geologists assume, so that their parents could take care of them. Scientists now think that many dinosaurs were **endothermic** (warm-blooded) and as social and intelligent as any animal of the Mesozoic era.

True mammals appear in the fossil record sometime in the Triassic period. During the Mesozoic, however, mammals remained small and apparently lived in the "cracks and crevices" left by other animals in the ecosystem.

Large animals learned to fly for the first time in the Mesozoic era. The pterosaurs were flying reptiles with hollow bones and may have been "gliders" rather than "flappers." Birds appear toward the end of the Jurassic period. The first birdlike animal that scientists know about is called Archaeopteryx and was discovered in fine-grained limestone, complete with fossil imprints of its feathers.

The Mesozoic era closes with the Cretaceous period, during which large chalk deposits were laid down in shallow marine environments, giving the period its Latinized name. Chalk is composed of the tiny skeletons of marine animals made of calcium carbonate. The Cretaceous period came to an abrupt end 65 million years ago with another mass extinction.

LIFE IN THE CENOZOIC ERA (65 MILLION YEARS AGO TO THE PRESENT)

Grasses first appeared in the Cenozoic era, and in parallel with that development, large grazing mammals flourished. Only very recently are primates and *Homo sapiens* noted in the Cenozoic fossil record.

The Cenozoic era has been divided up in several ways. The most modern nomenclature is as follows:

> Paleocene period: composed of the Paleocene, Eocene, and Oligocene epochs.
>
> Neogene period: Miocene, Pliocene, Pleistocene, and Holocene (or Recent) epochs.

But many geologists still use an older classification, namely:

> Tertiary: Paleocene through Pliocene epochs (from 65 million years ago to 2 million years ago).
>
> Quaternary: Pleistocene and Recent (the last 2 million years, informally known as the *Ice Age*).

The human family was present in the Miocene epoch, but the fossil record is sparse until the Pliocene. Although everything about humanity's history is hotly debated, *Homo erectus* appears to have been the first truly human animal, one that originated in Africa but spread to other continents. *Erectus* was a hunter-gatherer, made a number of stone tools, but left behind no evidence of abstract thought. At some point, for unknown reasons, *Erectus* became extinct.

Homo sapiens appears in the Pleistocene as the "Neanderthal man" (and woman). *Homo sapiens neanderthalensis* had, on average, a larger brain capacity than we modern humans do, and they also were huskier and stronger. Their culture was a more sophisticated Stone Age variety than what had been known before. Neanderthals disappeared from Europe about 35,000 years ago. Scientists now believe that *Homo sapiens sapiens* (an animal just like us) originated as a small population in Africa and coexisted for thousands of years with the Neanderthals. For unknown reasons, the Neanderthals became extinct, but we, *Homo sapiens sapiens*, have flourished.

"I'm afraid I don't have a name. They've only discovered about half the insect species."

What Can We Learn from Life's History?

SPECIES AND THE STRUCTURE OF TAXONOMY

The history of life on our planet makes clear several points. First, new species have repeatedly entered the fossil record. We can only marvel at the rich creativity that life on earth has displayed and the diversity of the current biosphere. On the other hand, the majority of species that once lived on land and sea no longer exist. For one reason or another, creativity on our planet has gone hand in hand with carnage. In the next chapter we will consider the evidence of extinctions in earth history and theories about the causes of mass extinctions. This chapter will accentuate the positive: the diversity of life and the historical record of increasingly complex life-forms.

One of the points about our diverse biosphere that has always impressed scientists is the discrete, distinctive nature of **species**. A species is a group of organisms that can breed to produce fertile offspring. That is, plants and animals come in particular kinds: Although we have both elephants and donkeys, we have no elephant-donkeys, much less elephant-donkeys with fertile offspring. There are rare exceptions to the discrete species rule, however: A few plants can be bred across species boundaries and produce fertile offspring, and for a few animals, biologists have trouble distinguishing between species and subspecies. But on the whole, one of the most remarkable aspects of life on our planet is that scientists can divide it into distinct types. Furthermore, this observation is not simply an artifact of our own science-obsessed culture: Indigenous populations around the world distinguish among species of birds and trees, and their naming systems generally correspond to biologists' divisions. For a century, biologists have argued about a number of theoretical points, but it is important to remember that the division of life into separate species is something scientists frequently agree on.

Biologists have also reached agreement about similarities among different species. That is, **taxonomists** (biologists who specialize in classifying life-forms) have successfully and consistently grouped species by degrees of similarity. An individual species belongs to a genus of similar species, and a genus is part of a family. Similar families are grouped together, and so on. Although one might imagine that life could be organized differently, species on this planet seem to exist in a manner lending itself to a hierarchical classification based on similar characteristics.[3] The traditional classification of humans, for example, is as follows:

> *Phylum: vertebrata.* All vertebrates have a backbone; those animals that lack backbones are put into another phylum.
>
> *Class: mammalia.* Mammals are vertebrates that bear live young, suckle their young, are warm-blooded (endothermic), and generally have body hair. Those kinds of vertebrates that do not have these features are put into other classes.
>
> *Order: primates.* Primates are mammals with opposable thumbs and include apes, chimpanzees, and monkeys. Other kinds of mammals are put into other orders.
>
> *Family: hominidae.* Members of this family are primates with bipedal locomotion and big brains. They were and are

[3]Above the level of species, the hierarchy is a human construction, in the sense that some families of plants do not have as large a range of morphological and genetic diversity as a family of insects does. Still, all biological scientists agree that a hierarchy of relatedness can be found among species.

relatively sophisticated animals, social and cooperative hunter-gatherers. They include *Australopithecus afarensis* and *Australopithecus africanus*, as well as the *Homo* species.

Genus: Homo. Homo erectus is the earliest member of this genus (appearing about 1.6 million years ago). *Homo erectus* had a larger brain than earlier species, and their culture appears to have been more highly developed. *Erectus* had many and varied stone tools. *Erectus's* teeth were smaller than those of *Australopithecus,* with a skull more modern in shape.

Species: sapiens. "Sapien" means wise: *Homo sapiens neanderthalensis,* known as "Neanderthal man" was the muscular "cave man" found in many popular representations of human evolution. It now appears that we did not descend from Neanderthals. Rather, *Homo sapiens sapiens* seems to have originated as a small population in Africa and spread across the globe. Neanderthals might be thought of as our "cousins" with whom we overlapped in time but who, for some reason, went extinct while we flourished.

WHY THIS ORDER?

Before the Darwinian revolution, biologists had no overarching theory that explained *why* life-forms were divided into species and *why* species shared characteristics in the hierarchical relationships of taxonomy. By any measure, this was a great gap in understanding. Scientists struggled with all the issues concerned with the relationships among species.

Some biologists thought that divine creation was responsible for life and that science could not determine why God had made the biological world in the way we find it. Such a situation, however, is far from satisfying for anyone wanting an intellectual explanation. Scientists, in particular, are taught to look for efficient causes of patterns in the natural world. Thus, in the nineteenth century, the scene was set for Darwin's spectacular entrance, our subject for the next chapter.

SUMMARY AND CONCLUSIONS

The history of life on earth is long and rich. Thanks to the fossil record, but always keeping in mind its imperfections and necessary biases, geologists now think that the earth's earliest life-forms were single-cell organisms living in the ocean. Complex, multicellular life first appeared in the Proterozoic eon.

The Cambrian period marks the "explosion" in the fossil record of animals with hard parts. In time, the vertebrates appear. Fish are followed by amphibians, which are followed by dinosaurs. Only in the Cenozoic era do mammals become dominant.

The fossil record shows repeated extinctions followed by the diversification of life-forms. The order in which plants and animals appear in the rock record gives us a glimpse of the long and complex history of life on this planet.

Species show a remarkable order, a natural and complex set of related characteristics. Taxonomists before the nineteenth century had no scientific theory that explained this order, but they nevertheless used the order to classify all plants and animals. An explanation of why these patterns of relationships were found throughout the living world awaited Charles Darwin and his intellectual descendants.

QUESTIONS FOR DEBATE AND REVIEW

1. What proportion of earth history came after the Cambrian explosion? Compare this figure, given as a percentage, with the proportion of earth history that came before the Cambrian.
2. What does Phanerozoic mean to a geologist? What do the Greek roots of the word mean? What are the Greek roots for Proterozoic, Paleozoic, Mesozoic, and Cenozoic? In what era did fish and amphibians appear in the fossil record? What era is dominated by dinosaur fossils? Birds? Mammals?
3. Why would the earliest life have been in the oceans rather than on land?
4. Give possible reasons for the regularities noted by taxonomists (the hierarchy of all living things).
5. Notice that the earliest forms of life, the simple bacteria and blue-green algae, still exist today. Also, consider some of the simplest animals of the early Paleozoic era, the brachiopods. They still exist, some in forms similar to their earliest fossil analogues. Why might relatively simple organisms have lasted through much longer periods of geologic time than did complex animals like dinosaurs?

CHAPTER 10

Darwin and Beyond: The Continuing Debate About the Origin of Species

The revolutionary intellectual changes unleashed by modern science may be said to begin with Galileo. As science has advanced in astronomy, physics, and finally biology, the traditional, religious interpretation of the natural world has retreated. The repercussions of the scientific revolution are still being digested in our society's collective spirit. Nowhere is this more true than in biology, in which evolutionary theories based on **naturalism**—or **materialism**—and studies of mass extinctions have challenged us to consider a self-sufficient, physical world regulated solely by efficient causality, a world with no underlying purpose and no necessary trends toward progress.

Terminology

Before we can even begin to discuss evolution, we face a obstacle that often makes students, scientists, and philosophers stumble. Both scientists and nonscientists use the word **evolution** for at least two different ideas, which are related but not identical. Only by carefully interpreting the context of the term can a reader understand a particular author's meaning.

1. *Evolution* can refer to the fact that the fossil record shows different species existing throughout the earth's history. That is, the history of life records great changes in both plants and animals. In the earliest rocks we see simple life. We know that during different eras one complex group of animals, such as the dinosaurs, were dominant. Only recently, geologically speaking, have mammals flourished.

2. *Evolution* can refer to the idea that the succession of species we see in the fossil record are related to one another through lineage. For example, fish gave rise to amphibians, and amphibians gave rise to reptiles. To confuse matters more, this notion is sometimes coupled with the idea of the mechanism or cause that led to such lineages, but sometimes it is not. When a cause is assumed, however, it usually is Darwinian natural selection, a concept we will discuss later in the chapter.

To make matters worse, some authors, especially nonscientists, equate a notion of progress—and not just change—with evolution. That is, in the popular culture, the idea of progress has been incorporated into our idea of evolution: The usual image is a succession of profiles of monkeys, chimps, Neanderthals, and, at the end of the sequence, a modern—almost always white, male. *Homo sapiens sapiens*. This standard image implies that we modern humans are the highest life-form on earth, the species that crowns evolutionary progress.

As you can see, the term *evolution* is used to refer to any one of a cluster of ideas. You must determine which sense is meant in a particular context. In the following chapters—and in newspaper articles, political speeches, and the rhetoric of school board meetings—you will have many opportunities to meet this challenge.

Darwin's Theory

THE UNCONTROVERSIAL PART

Charles Darwin, a nineteenth-century Englishman, focused his attention on the diversity of plants and animals and on the geologic history of life as it was then known. On his famous trip around the world on the English ship *Beagle*, from 1831 to 1836, he traveled throughout the Southern Hemisphere describing the local geology and collecting biological specimens of the native flora and fauna. Although Darwin was a young and inexperienced man at the time, he nevertheless took extensive notes on the similarities and differences of the many animal and plant species he encountered. He toured small and isolated islands like the Galapagos where he saw unique animal populations that he studied and sketched. Darwin noted that on each of the Galapagos islands, the birds and lizards differed, even though the islands were not very far apart and all of them probably were populated by animals from the South American continent, also not too terribly far away. In addition, Darwin was interested in geology as well as biology,* and so he collected rock samples, including those containing fossils, from around the world.

After returning to England, Darwin developed a theory to explain the biological diversity he had cataloged in living species and the differences he noticed between certain living animals and their closest fossil analogues. We will discuss this revolutionary theory later in the chapter, but first we will consider those points that Darwin's thinking had in common with that of other naturalists of his time.

Darwin and his contemporaries all knew that horse breeders, cattle ranchers, and dog trainers shaped the characteristics of their animals simply by selecting for breeding those with desirable traits. This man-made selection, operating for many generations, had created toy dogs and sturdy sheep dogs; it had produced delicate-boned racehorses and hefty workhorses. Darwin's general observation of this diversity of domestic stock was not a controversial contribution to biological studies. But the power of selectivity in breeding domestic plants and animals had not been specifically articulated before Darwin, who, remember, formulated his theory before the basic laws of genetics had been widely understood. Today scientists call this kind of change within a species **microevolution.**

Such biological change might be considered the narrowest and least controversial part of Darwin's evolutionary theory. It is accepted today even by "creation scientists," or biblical literalists, who focus their criticism of Darwinism on the generation of new species, not on the alteration of characteristics within a species.

Microevolution can easily be demonstrated experimentally: Bacteria, cultured in laboratories, may supply the population to be tested. When penicillin is added to the environment in which a single species of bacteria is living, most individuals die out. But those few bacteria resistant to penicillin survive and reproduce. Eventually the population is composed of the same species of bacteria, though one possessing a new characteristic, namely, resistance to the antibiotic agent.

*Darwin's university degree was earned in geology—a fact modern biologists have forgotten when they claim him for their own.

Notice that microevolution by itself is not an overarching theory explaining the similarities of related species. But Darwin used his observations about the wide variation of domestic animals as a springboard for more fundamental theorizing.

Another pair of ideas that Darwin explored after his return to England came from the part of biology called *comparative anatomy*, the study of animals' body parts. There were two points that nineteenth-century biologists agreed on:

1. Some organisms have vestigial structures that are small and functionless. Some birds, for example, have stubby wings that are useless for flight. Or to take another example, we humans have a true tailbone but no tail.

2. Anatomical analogies can be detected among most organisms, even though the body parts in question perform different functions. For example, the wings of a bat, the flippers of a seal, and the arms of a monkey all are made of the same grouping of bones. These sets of bones are called *homologous structures*.

The observations of comparative anatomy were not, in themselves, controversial. Instead, Darwin's genius lay in his interpretation of the observations that many had made before him.[1]

THE CONTROVERSIAL PART

Darwin's broader theory (known as **Darwinism**), first published in *The Origin of Species* in 1859, is sufficiently subtle that it has been presented differently by different writers, both those favorably disposed to it and those who, for various reasons, criticize it. The following is a summary of the theory's most important points.

Darwin's observations convinced him that not only could individuals within a species vary *but also species themselves could change and generate multiple new species over time.* This idea is represented by the term **macroevolution**. The beauty of this idea is that it explains why species are similar, why taxonomic

[1]Another English naturalist, Alfred Russel Wallace, also gathered evidence and conceived ideas similar to Darwin's. Although Wallace did not have Darwin's comfortable middle-class background and worked more independently of his fellow naturalists, far from England, he nevertheless developed all the major ideas usually credited to Darwin. In the end, Wallace's work helped persuade Darwin to move forward and publish his work, which Darwin anticipated would attract the hostile attentions of scientific and religious communities around the world.

divisions are neatly hierarchical, and why organisms have homologous and vestigial structures. Species are like one another, Darwin proposed, because they are *related through historical events that link their lineages.* Taxonomic classifications merely reflect the ultimate parentage of all species. Vestigial structures are elegantly explained by the notion that the useless wings of flightless birds are modified appendages inherited from early birds who did fly. Similarly, homologous structures in different species exist because the organisms that possess them are descended from a common ancestor. Darwin summed up these ideas in the phrase **"descent with modification."**

Because of its emphasis on common lineage, Darwinism is an overarching theory, embracing biologists' and taxonomists' most important observations and giving explanatory power to a previously descriptive science. Although controversial from the outset, this theory appealed to many scientists because it was so powerful, linking previously separate data and ideas. It was, indeed, a creative contribution to biological theory unparalleled by any other theory in its era.

Darwinism stated that similar species are related because they have evolved, one from another. What mechanism, Darwin asked, leads to such change and to the generation of new species? He argued that the key was *natural selection operating on variation within a population.* He believed that species gradually changed because those individuals that had won the struggle to survive and reproduce passed on their characteristics to the next generation. In a world dominated by "survival of the fittest," a species would gradually change as novel and beneficial characteristics—produced randomly—replaced other characteristics of the population. Darwin believed that over many, many generations, natural selection gradually led to the generation of a new species from the old. One way of summarizing Darwin's view is to say that in general, species do not actually go extinct in the sense of disappearing; they just evolve into a new, and better adapted, species. This occurs gradually as each new species appears and the old one fades away.

PROBLEMS WITH DARWINISM AND DARWIN'S OWN RESPONSE

Darwin himself recognized that there were problems with his theory. In his first book, he discussed the rarity of transitional fossil species. Because he argued that a new species was a result of gradual changes in an existing species, Darwin thought that step-by-step changes in species should be evident in the fossil record. But in fact, the fossils known to Darwin and his colleagues did not fit this pattern. Instead, they showed the general absence of intermediate species. Darwin dealt with this problem in his book by contending that the rarity of all preserved fossils accounted for the "gaps" in the fossil record. Geologists have, he believed, only a few "snapshots" of the history of life, and therefore it is not surprising that changes seem sudden rather than gradual.

Another problem that Darwin noted was the difficulty one might envision in creating, in a series of gradual steps, complex organs like the eye. What selective advantage, a scientist might ask, could the first few "eye cells" have? That is, why should a mutation that produces a few extra cells be an advantage to the individual organism carrying this mutation? But such advantages are necessary for Darwin's theory at each and every gradual step of evolution. In response to this sticky question, Darwin noted that the telescope, a manufactured instrument, took generations to perfect through a series of designs.[2] He argued that in an analogous manner, the eyeball had evolved through time, with each intermediate stage useful to the animal that possessed it. This was true of even the most primitive light-receptive cells, for a tiny bit of vision is better than no vision. Gradual improvements in the eye thus were favored by natural selection, since in most animals, better vision favors survival and the likelihood of reproduction.

GRADUALISM AGAIN

Gradualism was important to Darwin because it was part of the orthodox scientific view of the nineteenth century. Darwin was a close friend of the great geologist Charles Lyell, whose influential book on gradualist geological principles Darwin read aboard the *Beagle*. Darwin was, as a matter of methodological principle, as committed to gradualism as Lyell was. Accordingly, this aspect of Darwin's theory was not controversial within the scientific establishment of his day, in part because it was viewed as a methodological improvement on all previous biological theory. Biologists before Darwin thought that species were suddenly generated in steps or jumps. No matter what the cause of such events, natural or supernatural, the emphasis had been on the sudden appearance of new species. But sudden events buried deep in the past of geologic history are much more

[2]Notice that Darwin's choice for an analogy puts him on weak ground. The telescope has "evolved" because of human *will, choice,* and the *goal* of better and better long-distance vision. This is a final cause. Darwinism is correct only if some efficient cause leads to the changes that Darwin's theory predicts will occur in small steps over long time periods.

difficult to investigate scientifically than are gradual changes. Thus to most scientists, Darwin's hypothesis seemed to be a great intellectual leap forward.

DARWINISM RAISES NEW QUESTIONS

Variations within domestic animals had been the starting point for Darwin's reasoning. A natural question at this point is, Have we humans created new species with our selectivity regarding breeding animals? Unfortunately, when scientists examine this idea in detail, the boundary line of exactly what constitutes a species can be fuzzy enough that they are not sure quite what we have invented. Biologists say that species differ from one another when they do not interbreed under natural conditions. Have we humans created any new species in our domestic breeding programs?

Take the example of the beloved family dog. We humans have controlled—that is, selected—the breeding of canines for about 14,000 years. The many different types of dogs we have created through selective breeding are termed *breeds* by dog lovers or *strains* by scientists. The many breeds of dogs are astoundingly different, but they can and do breed across the distinctions so carefully spelled out by the American Kennel Club. In short, dog breeds are not dog species.

But are all dogs, as a group, even a species? Despite a 14,000-year history of "selection" controlled by humans, some domestic dogs can and do interbreed with coyotes. Other dogs can and do interbreed with wolves. It would seem, therefore, that dogs, coyotes, and wolves should be one species, even though biologists consider them as members of three species. This is clearly a matter of judgment, and any attempt to draw a sharp line between certain species can be criticized from several standpoints.

In general, Darwin and his followers did not claim that domestic animals were an important example of the creation of new species by means of careful selection. Critics can point to this as a possible weakness in the theory, as indeed, "creation scientists" still do today.

THE GENERAL DEBATE

But we have gotten ahead of ourselves. We need to go back to the nineteenth century and consider the range and depth of the debate regarding Darwin's theory. One of the revolutionary characteristics of Darwin's theory of natural selection was that it reflected purely naturalistic and efficient causality. Darwinism is an example of **philosophical materialism** or **naturalism**.

That is, the theory does not appeal to the supernatural or immaterial but instead explains data with reference only to matter and the efficient causality of the natural world.

Even in the case of human evolution, Darwin clung to materialism, rejecting such long-established views as philosophical **dualism,** popular with most nineteenth-century intellectuals. Dualists link matter (for example, the human brain) and nonmaterial spirit (the mind). But according to Darwin, there was no spirit and no hand of God recorded in the fossil record. The evolution of life, Darwin was sure, could be explained solely by natural selection or other efficient causes. For the first time, scientists had a grand theory that did not depend on divine will or any other final cause to explain the evidence that geologists found in the fossil record and biologists saw in taxonomy and anatomy.

Darwin's theory directly contradicted the popular nineteenth-century philosophical or theological argument, the "argument from design." Many English thinkers pondered the implications of the complexity and interrelation of the biological and physical worlds, which, they believed, were indications of the work of a designer, a creator God. Darwin's publications offered an impressively detailed explanation of this biological complexity, which, however, depended only on efficient causation, with no room for any sort of designer.

The spirit of Darwin's theory gave paleontology and biology new strength, but it also placed them at the center of social controversy. Giving a wholly materialistic explanation for the variety of life on earth and for the existence of our own species makes the world look like a kind of machine, which causes social and religious conservatives to argue that Darwinism excludes more than a Divine Being from the earth. There seems little or no room for any kind of spirit, for any sort of final causes, or for human free will or the mind or morality. Indeed, *Darwin's picture of the world as a machine* has alarmed many progressive social leaders and even some scientists.

Moreover, and more disturbing to almost everyone, Darwin did not believe that biological evolution on earth was progressive. That is, he rejected the notion that there were "higher" and "lower" species and that humans stood at the top of the ladder of evolutionary progress. Evolution, Darwin was certain, simply worked to adapt animals to their local environments, and so *there was no reason to expect "progress" from such a system.* This aspect of Darwinism distinguished it from other, competing, views of evolution debated in nineteenth-century scientific circles.

Finally, yet another aspect of Darwinism appalled some people. Thinkers from Aristotle onward had believed that *Homo sapiens sapiens* was created separately from other animals and, owing to its **separate creation,** possessed distinctive fundamental characteristics. Darwin's theory undermined that view, and in 1871, he published a book on the origin of humankind that explicitly stated that humans descended, with modifications, from earlier primates. Many church members of Darwin's day took strong exception to this aspect of his theory.

THE TWENTIETH CENTURY

If Copernicus removed the earth from the center of the universe, Darwin removed humankind from the center of earthly creation. He also removed the idea of a Divine Being from the long and rich record of the history of life. No God-given or Aristotelian final causes were needed, according to Darwinism, to explain the wonderful diversity of life and its long evolutionary history.

Most biologists and paleontologists had good reason to accept Darwin's theory. As just one example, consider the unusual species of animals and plants living on isolated islands. From the Galapagos to Hawaii to New Zealand, Darwin and other European travelers had cataloged truly unique species. But traditional biology could not explain such unusual creatures as the kiwi bird of New Zealand or the one-of-a-kind fruit flies of Hawaii or the giant tortoises of the Galapagos. The power of Darwin's theory was shown by these unusual creatures because Darwinism could claim to explain their origins. According to Darwinism, the small and isolated populations of island animals are especially susceptible to the effects of variations of what we now call **mutation** events. In large mainland populations, one individual with a favorable characteristic has little impact on the makeup of the whole population. But in small populations, such as those on isolated Pacific islands, a single unusual individual can make a significant difference in what its descendants will inherit. An adherent to Darwinism thus can claim that favorable characteristics more rapidly change an isolated species. And it was precisely this, we might reason, that Europeans had discovered and described throughout the British Empire.

In short, in this century it has seemed that although evolutionary theory appears to have some weak spots, as Darwin himself noted, natural selection has been able to explain phenomena that previously had only been described. Like plate tectonics, Darwinism is an overarching theory under which smaller theories and different types of observations can be nestled.

In recent years, George Gaylord Simpson, Ernst Mayr, Edward O. Wilson, and other eminent biologists have developed modifications of Darwinism in light of more advanced work concerning genetics and the selection process itself. These scientists are often termed **neo-Darwinists,** and in recent years their work has dominated orthodox biological research.

But the gap between the gradualism of Darwinism and neo-Darwinism, on the one hand, and the fossil record, on the other, has become increasingly clear to geologists as this century has progressed. Diligent scientists have collected tens of thousands of samples of species unknown to Darwin, but they have not found all the transitional species that Darwinism and neo-Darwinism seem fully to expect. Indeed, in many or even most parts of the fossil record, there is no gradual history leading to a new species, whereas we frequently see some species spring unannounced into the fossil record.

Recently, some geologists have concluded that the essence of Darwin's theory can stand without such rigid insistence on gradualism.[3] A mechanism for the relatively rapid generation of new species is needed to keep natural selection in accord with some of the abrupt changes we see in the fossil record. Modern theorists have advanced ideas about such rapidity under the term **punctuated equilibrium.** They note that most of the evolutionary changes recorded in the fossil record occurred not during the long life of a species but in the short periods of time during which a new species was generated. In 1972 Niles Eldredge and Stephen Jay Gould jointly proposed punctuated equilibrium as a theory that better fit the fossil record. Eldredge had studied Paleozoic trilobites and found exactly what is now taught to introductory geology students everywhere: Most trilobite species show fairly constant traits through time, but new species suddenly appeared and disappeared throughout the record of the Paleozoic eon, making many trilobites good "index fossils."

Eldredge and Gould labeled as stasis the unchanging qualities of most species through time. In their view, rapid evolutionary change occurs during short bursts of speciation, that is, the relatively short period of time when a new species comes into being. These short but dramatic phases "punctuate" the "equilibrium" represented in the static periods.

Eldredge and Gould believe that the geologically sudden changes inherent in the speciation event depend on the isolation of a small population of the original species. Significant mutations in a small population can substantially affect the gene pool. If the mutations are selectively advantageous, the theory goes, they can lead to a geologically rapid change in the organism, indeed to the generation of a new species. The key here is the role of selection: The

[3]The possibility of the rapid generation of new species was explored by neo-Darwinist biologists like Ernst Mayr. But it is the geologists, because of their concern with the fossil record, who have most emphasized such change.

isolated population—for example, animals on an island—can be exposed to selective pressures different from more global ones. The adaptations that are selected can therefore tailor the organism to a biological **niche** somewhat different from that of the parent organism. If the new species flourishes in its previously unoccupied biological niche, one might expect it to become widespread quickly. Thus the new species appears abruptly in the fossil record in many different localities (Figure 10-1).

Notice, once more, that the Eldredge–Gould theory calls for most evolutionary change to occur within speciation events, not through the gradual modification of existing species. This is in accord with the fossil record, but it places an increased burden on speciation, an event for which we have little evidence and, some might say, that we understand only dimly.

From a neo-Darwinist, gradualist perspective, punctuated equilibrium theory runs the risk of relying on mutations so substantial they border on the miraculous "jumps" of earlier biological theory. Is it reasonable, skeptics ask, for mutations to be big enough to create a new species in 100 generations? If so, why don't we see similar speciation events in the laboratory when we breed thousands of generations of bacteria and fruit flies or, again, in our long record of breeding domestic animals?

It is important to remember that *punctuated equilibrium theory was developed by paleontologists, not biologists.* Thus it is the fossil record to which Eldredge and Gould refer, and by their very nature, fossils record only

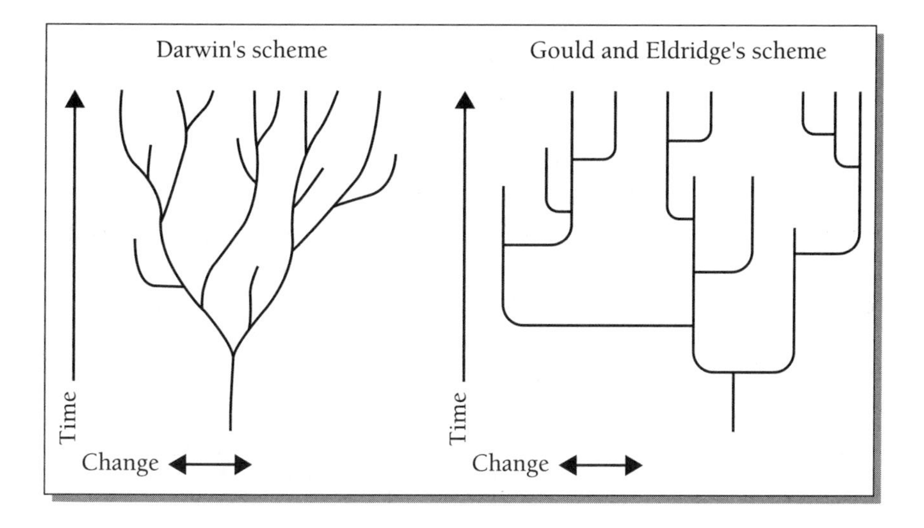

FIGURE 10-1: Theories of evolution.

morphological (shape) changes. Fossils contain little information about the interaction of species, their behavior and cooperation, predator/prey relationships, and other matters crucial to the construction of a biological niche. In living organisms, biologists can study behavior, the life history of an individual, and genetic variation that is not expressed morphologically, whereas geologists can examine only static fossils that disclose very little information about an organism other than its basic size and shape.[4]

The strengths and weaknesses of punctuated equilibrium continue to be debated in paleontological and biological circles. The discussion is partly just a reflection of the larger debate in natural science about the rate of change in earth history. Clearly, gradualism in paleontology and evolutionary biology can no longer be assumed; gradual change must be demonstrated. Although geologists are unsure about many things, it does seem certain that their methodological framework has become more complex than it was in the day of Darwin and his geological mentor, Lyell.

Challenges to All Forms of Darwinism

METHODOLOGY ONCE AGAIN

Contrary to the assumptions of most evolutionary biologists, large philosophical differences remain in the broader intellectual community concerning the development of life throughout the earth's history. The core of evolutionary theory hinges on one of the methodological assumptions of science, namely, efficient causality. Remember that an efficient cause is an immediately preceding and natural event, used to explain rationally what follows it.

Although scientists in chemistry and physics accepted efficient causes for natural phenomena, biologists before Darwin resisted them. Organisms, after all, seem to develop toward an adult state as if it were their natural goal, and progress toward a goal is part of a framework of final, not efficient, causality. Higher animals, it was pointed out, make decisions. It therefore often makes sense to credit them with purpose, and that clearly accords with final causality dependent on future states.

[4]Another shortcoming of paleontology is that it studies only the diversity of individual species. Biologists point out that the interaction among species is crucial to their survival. This interaction influences changes in multiple species through a process known as *coevolution*. The fossil record, once again, is mute on this point, since it records very little about an organism's behavior and ecological relationships.

Darwin and his intellectual successors appeal only to efficient causality to explain all biological phenomenon. Thus, all questions about life on our planet become problems that science can address. Indeed, Darwin's basic theory was so successful at organizing what scientists know about the varieties of life that biologists dropped their earlier assumptions about goals and purpose. It was only with the triumph of Darwinism that final causes and overarching "goals" were eliminated from standard biological science. But whether this belated change was wise is unclear to at least some intellectuals.

Another methodological problem is built into Darwinism. There is a redundant loop in his argument:

- Who are the fittest individuals? Only those who survive and reproduce.
- Why do they survive and reproduce? Because they are the fittest.

Or again:

- Why did turtles survive the Cretaceous extinction (which apparently wiped out many marine organisms and, quite possibly, the dinosaurs)? Because turtles were more fit than some other species.
- Why were they more fit? Because they survived.

Although natural selection does look like a very important part of what must be happening in the history of life, Darwinists often express themselves in a way that explains anything that exists. This, of course, is tantamount to explaining nothing.

Other Questions Yet to Be Answered

In this century a few scientists and other intellectuals have found the driving mechanism of Darwinian evolution to be implausible. One part of the problem is summed up by the questions, Can the extreme changes in life that geologists have carefully cataloged in the fossil record be driven by random mutations even if they are sorted by natural selection? How can complex adaptive features, especially those that involve new and fully formed structures, be created solely by the action of chance and elimination? Important to us humans, why are nonadaptive capabilities (like musical talent and mathematical creativity) thrown into the human population by means of mutation? After all, our finest talents do not seem related to the survival of the species or to our chances for reproduction. Some other scientists note that on average, random change in a system leads to lower states of order, not higher ones. It seems counterintuitive to some thinkers that a random change in precious genetic information, even a random change shaped by natural selection, would lead, over and over again, to more complex patterns of life.

But biologists remain firmly committed to Darwin's basic ideas and contend that their nonscientific colleagues overlook the creative power inherent in the process of natural selection. It is wrong, they argue, to emphasize the chance nature of mutation when what matters is *which* mutations are selected to be preserved in the next generation.

Another difficulty, however, might be mentioned here. Darwinism and its intellectual descendants hold that individuals strive to reproduce, to pass on their characteristics to the next generation. Modern biochemistry has shown us the details of genetic inheritance in the DNA molecule. But traditional evolutionary theory seems to face a fundamental problem: the existence, indeed the predominance, of sexual reproduction. If the driving force of evolution is passing along an individual's genes to the next generation, asexual reproduction would seem to be preferable to sexual reproduction, in which an individual sees only 50 percent of its genes represented in an offspring. But higher organisms, including all mammals, reproduce only by sexual means. *Why should this conundrum exist in a world driven by Darwinian forces?* This and other questions have yet to be resolved by biologists, a fact that leaves other thinkers room for some free-ranging speculation.

A few unorthodox, academics argue that evolution is driven toward higher and more complex life-forms. There is a long-term and clear trend toward consciousness in the fossil record, they say, that cannot be explained solely by the efficient causality of natural selection. The drive toward complexity seems to indicate something contrary to the spirit of the second law of thermodynamics, something that looks very much indeed like progress. This is true especially because complexity has led to "higher" things like art and knowledge and the emergent phenomena of consciousness, purposiveness, and morality. Human community reflects properties simply not present in a bee colony or an anthill. This type of argument was articulated in the first part of this century by the French Jesuit and scientist Pierre Teilhard de Chardin and has been taken up by others since then.

More recently, a few people have speculated that evolution is driven to fill all possible niches in an ecosystem. After all, if a niche is not occupied, anything that stumbles or steps into it will probably thrive. Thus more and more niches become occupied through time. From this viewpoint, filling all possible biological niches is the state toward which evolution is taking the biosphere, rather like entropy driving inorganic systems toward their lowest possible energy states.

In short, highly different types of theories are being advanced by a few people today in response to the apparent weakness of placing so much stress on random genetic change and natural selection as the engines of evolution. But traditional defenders of Darwin's theories, both neo-Darwinists and enthusiasts for punctuated equilibrium, strongly criticize these developments.

SUMMARY AND CONCLUSIONS

Darwin's theory helps scientists organize and explain the diversity of life, the hierarchical similarities of different species, and homologous and vestigial anatomical structures. Darwinism's strength lies in the fact that it uses only fully naturalistic and efficient causality to explain what had previously been considered to be outside the realm of empirical research. But Darwinism was and is still rejected by many members of our society.

Recent scientific research has emphasized that Darwinian evolution does not require tiny incremental changes. Darwin was committed to gradualism because of the intellectual climate of his age. But modern scientists are impressed by the fact that most species in the fossil record seem to remain static for long periods of time, with new species generated in short bursts of creativity. The generation of new species is central to evolution, and from all views, it is not fully understood. Better information or theories about speciation would be most helpful in the current debates in established scientific circles.

Some writers argue that random genetic change cannot lead to the complex adaptive features we see around us. But most paleontologists and evolutionary biologists are still firmly in the Darwinian camp and defend the idea that natural selection acting on random variations is at least the *main* engine for evolutionary change. After all, these scientists contend, the hallmark of a good theory is the width of inquiry to which it can satisfactorily respond. Darwinism is powerful: What else could replace it and still be as scientifically fruitful?

QUESTIONS FOR DEBATE AND REVIEW

1. Darwin lies buried in Westminster Abbey, the highest honor the Church of England can grant a scientist. Does Darwin's resting place surprise you? Explain. Why might the Church of England have been glad to receive him?

2. Does materialism in itself mean that the history of life must not be progressive, as Darwin claimed? Explain, perhaps with reference to Adam Smith's invisible hand.
3. Does the overall history of life sketched in the last chapter look nonprogressive? Do you think that "higher species" and "lower species" are terms that have meaning in the natural world?
4. Describe Darwin's view of natural selection and its effects.

CHAPTER 11

Mass Extinctions: Triumph for Catastrophism?

Only about one in 1000 species that existed in the past is still alive today. Thus, when a geologist picks up a fossil in the field, it is highly unlikely that the species still has living counterparts. Geologists study this impressive global death rate of species because the fossil record, and the extinctions it records, gives them a timescale for earth history. In other words, geologists cheerfully construct the earth's lifetime calendar based on the mortality of species.

Extinctions in Earth History

SLOW VERSUS RAPID RATES OF EXTINCTION

In general, extinction need not have any surprising or dramatic cause. If a species is increasingly limited geographically, perhaps because of changing climatic patterns, it will become isolated from other regions in which it might have flourished. The species' population base will thus grow smaller as its geographic distribution shrinks. The individuals may be further reduced by an epidemic disease, by a sudden loss of habitat (for example, because of a forest fire), or by the extinction of another species that had served as prey. In short, we think that a species can become extinct under quite normal circumstances. Such events are known as **background**

extinctions and are found in all parts of the fossil record. Such extinctions also fit into the framework of gradualism. But some extinctions appear to affect simultaneously a number of different species, all perishing at the same time. This, of course, has the distinct odor of catastrophism.

At least at first glance, scientists can see why some extinctions might lead rapidly to others. In the dependency of the food web, the extinction of a lower species may well trigger difficulties for the survival of many higher species. In short, the dreaded "domino effect" may influence what happens during major extinction events.

DIRECTIONALITY AND EXTINCTION

Human nature seems drawn to a search for meaning, or at least direction, in the world around us. Most of us do not like to rest our lives on the ideas associated with random death. Perhaps it is not surprising, therefore, that both Darwinian scientists and many laypeople are happy to interpret earth history in a manner that awards some utility to extinction.

Many scientists unconsciously assume that the "best-adapted" species survive the continuing and global competitive struggle for limited habitat. These successful species, scientists believe, contribute to new species that can prosper in new ecological niches. Such a view has a direct impact on the scientific understanding of human origins. Scientists know that several species of hominids existed before *Homo sapiens*, and there is evidence that two of these species probably overlapped in time. As humans, scientists would perhaps like to think that there was something about the earlier species that was not as "good" as the later ones. We exist, scientists often unconsciously assume, because we are so clever, because our talents have earned us the right to be here.

Unfortunately for our emotional comfort, there is good geologic evidence that extinctions are quite random. It would appear that there is such a thing as being in the wrong place at the wrong time. Those who study the fossil record believe that some long-lived and seemingly well adapted species lose the struggle for survival more or less by chance. It seems that "bad luck" has more to do with the demise of most species than do "bad genes" or maladaption, although the matter is far from settled.

DIRECTIONLESS BUT NEVERTHELESS USEFUL?

Even if most extinctions are random, wiping out perfectly good species, could it be that extinction itself has uses? Some geologists and evolutionary biologists have argued that new ecological possibilities are created by extinctions. The demise of the dinosaurs and many large reptiles at the end of the Mesozoic era, for example, may have been necessary to clear "space" for mammals to diversify and flourish in the Cenozoic era.

Some researchers believe that evolution without extinction would mean that all organisms would rapidly diversify, without ever stopping. Without extinctions, new species would accumulate everywhere on the planet. The generation of new species, the event we call *speciation*, might grind to a halt because there would be no niches left for new species to exploit.

But in any event, it is clear that for better or worse, extinctions occur. Both background extinctions and more major events have shaped life's history.

THE CONCEPT OF MASS EXTINCTIONS

Many geologists have recently turned their attention to periods in earth history during which many species and genera of life have ceased to exist at approximately the same time. The public has been fascinated in the last decade by the concept of **mass extinction.** From television programs to comic strips, you have no doubt seen reference to sudden extinctions, especially the event proposed for the demise of the dinosaurs.

Geologists generally recognize five episodes of mass extinction in the Phanerozoic eon, often called the *Big Five*. Their names are based on when they occurred on the geologic timescale (see Appendix C). In order of decreasing size they are as follows:

- End of the Permian. This was the largest extinction in earth history, ending the Paleozoic era. This was the time when trilobites and many others met their demise.
- End of the Cretaceous. This was the extinction that may have swept the dinosaurs off the earth.
- End of the Triassic. This was a sharp interruption in the dinosaurs' reign, but not the end of the Mesozoic story.
- Near the end of the Devonian. This is the only one of the Big Five not exactly on a period boundary.
- End of the Ordovician. This extinction was the smallest of the Big Five.

There is an important debate concerning the Big Five, however, that has not reached the general public. Were they real events? That is, are mass extinctions different in kind from normal, ongoing extinctions of individual species? It could be that mass extinctions are simply more extreme cases of normal and continuous **background extinctions.** From this point of view, the Big Five are indeed unusual, but then unusual events are to be expected every now and then in the millions of years of earth history. To put it another way, mass extinctions may be like 100-year floods: They are larger in magnitude than more common and smaller flood events, but they are not caused by fundamentally different processes.

Two researchers at the University of Chicago studied the rate of extinction throughout the Phanerozoic eon. Using statistical analysis, they concluded that the Big Five extinctions were sufficiently greater than others to merit a special status.[1] But the data are few, and more information from the existing fossil record, or independent information about poorly preserved species, could change their calculation one way or another.[2] Figure 11-1 is a histogram from another work showing an apparently smooth trend that would not lead one to identify the Big Five as special events.[3]

Nevertheless, the unique nature of mass extinction is credible to many geologists. It seems to have been accepted without question by physicists who recently, and quite dramatically, have studied earth history, a point to which we will return in a moment. Note that the geologists who recognize mass extinctions as unique events quite different from background extinctions, are necessarily sympathetic to the possibility of catastrophic change in earth history. Such a view adds importance to the concept of contingency, or dependence on some prior event, in the history of life. This fact gives yet another emotional twist to debates about extinctions.

Broader Implications of Extinctions

CONTINGENCY AND OUR EXISTENCE

Geologists believe that the history of the physical earth and the history of life on our planet are intimately interrelated, that each has directly influenced the other and neither would be what it is today except for these interactions.

[1]D. M. Raup and J. J. Sepkoski, "Mass Extinctions in the Marine Fossil Record," *Science* 215 (1982): 1501–1503.

[2]See, for example, M. J. Benton, "Diversification and Extinction in the History of Life," *Science*, April 7, 1995, pp. 52–58.

[3]David M. Raup, *Extinction: Bad Genes or Bad Luck?* (New York: Norton, 1991), p. 81.

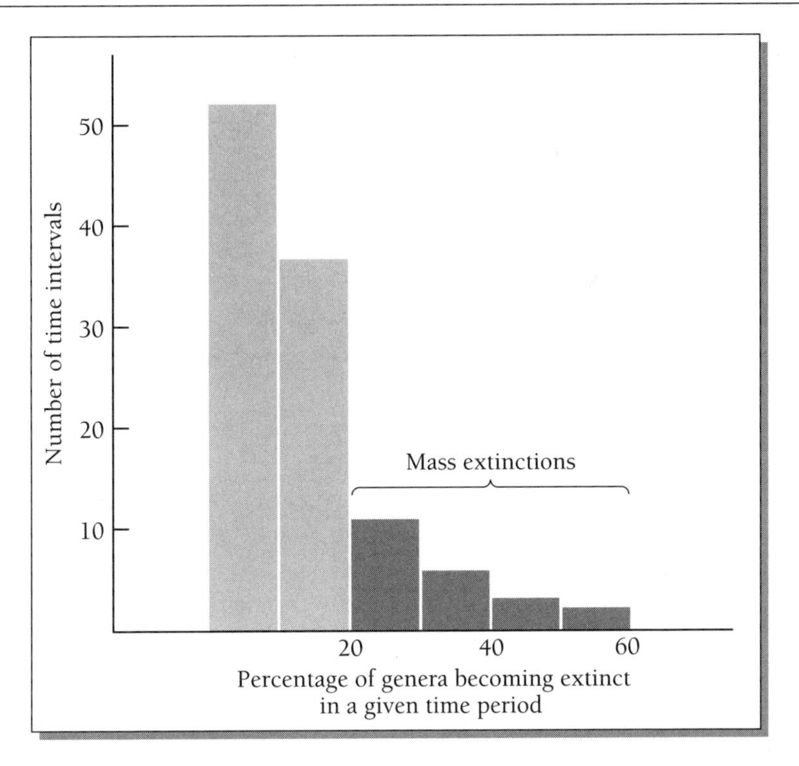

FIGURE 11-1: Phanerozoic genera of marine animals (106 time intervals = total). This figure shows a smooth change from numerous "background" extinctions. This suggests that special causes of mass extinctions may not be an appropriate model. From David M. Raup (1991).

Furthermore, the enormous length of our planet's history and the complexity of biological and physical interactions indicate that the story of the earth could have turned out quite differently. If we could go back in time to the formation of the earth and allow history to start again, geologists believe that a different story would emerge. Even if life originated in some similar fashion (a big "if"), complex species very different from what we know might dominate the planet. Small changes in the history of the earth, even the survival of an individual amphibian or mollusk, could preserve a species from extinction. This, in turn, could alter the next stage of life's evolution. In other words, each event in earth history is contingent on small differences in preceding events.

Geologists can see no reason to suppose the succession of plants and animals that we find in the fossil record had to come into being as they did. The Ediacara fauna might have flourished indefinitely. Or again, mammals might or might not have existed in the Mesozoic era. If they did, they might well have become extinct when the dinosaurs met their demise. From this perspective, it seems most improbable that *Homo sapiens sapiens* would come into being a second time around. If we started earth history over again, the odds for any kind of hominid would probably be much less than one in a zillion.

A PAUSE TO THINK

Scientists may be entirely correct about the earth's history, but the notion of contingency is worth careful consideration. Geologists have been known to confuse methods of inquiry with substantive claims about the world. That is, geologists sometimes assume that because they inquire about things in a certain way, the world must indeed be that way.

One can argue that contingency is incoherent in the discipline of science, at least if it is taken as a substantive claim about the world. It is one thing to say that for purposes of inquiry, geologists assume that natural *events* are contingent. But it is another matter altogether to declare flatly that the natural *order* is contingent. A philosopher might point out that such a declaration presupposes some point of reference existing outside the natural order that is itself not contingent. Otherwise, the claim to contingency falls into a problem of self-reference. In order to avoid saying "Everything is contingent (including this statement)," one must say, "Everything is contingent, except this statement and some other things (whatever else one might want to exempt)." Otherwise, logically, it is like trying to push a cart and ride in it at the same time.

Some philosophers argue that in order to have a universe of contingent things, we must presuppose a noncontingent something, and whatever that is, it must also be a "necessary being." Religious readers may recognize this as one form of Thomas Aquinas's proofs of God's existence. But note that logic demands that this God stands outside the natural system, a wholly other kind of Being. This is quite in contrast to the picture of a God who interacts with the world as often sketched by clergy in the pulpit.

AN EVEN BIGGER PICTURE

Current philosophical debate has focused on contingency in more than earth history. The apparent improbability of the life-forms on our planet can be more than matched by the improbability of the universe itself. Why this universe and not some other? As science has shown, the improbability of our

universe can be expressed with reference to the ratios between certain crucial physical forces. Such ratios are known as *natural constants*, for example, the ideal gas constant encountered in high school chemistry.

Consider for a moment a more fundamental ratio: The exploding force of the Big Bang would have to have been balanced against the implosive force expressed in the law of gravity, in such a way not only to prevent an immediate collapse of the mass of the universe back onto itself, but also to allow the universe to evolve outward in the way it has. To take another example, consider the forces within a star of the habitable variety, that is, a star like our sun, which has a stability permitting life nearby. The existence of such a star depends on a ratio called the *gravitational coupling constant*. This ratio balances the exploding force of the nuclear reactions within a star against the force of gravity. Thus the gravitional coupling constant of a habitable star prevents a star's evolution from being mostly that of a blue giant (which dies too quickly to give life a chance to develop) or a red dwarf (which radiates too weakly to keep water above freezing on nearby planets, a requirement for all life as we know it).

It seems quite possible that natural constants could have had other values than they do. That they are what they are is what has made our universe possible. We might take the view that the universe has these ratios by chance. It may be improbable that all the ratios have worked out to our advantage, but if they had not, we would not be here. However, some philosophers (and even some scientists) have speculated that the natural constants on which our existence seems to be contingent result from some sort of final cause acted out through "fine tuning" of the universe. This idea is known as the **anthropic principle**. According to this view, it is because we *Homo sapiens sapiens* are here (despite the staggering odds against us) that we are led to suspect a final cause is at work, that the universe has a goal or direction of which we are at least one part. Religious thinkers enter at this point and construct arguments for the existence of a God based on the improbable "design" of the world.

In short, both scientists and nonscientists would be well served by remembering that as a discipline, science assumes the efficient causality of all events in the natural world. Science cannot, by itself, address any questions related to final causes. Such "intentional" causality may or may not be central to our world and its history of repeated extinctions, but it remains an active part of the thinking of many members of our society.

The Current Scientific Debate About Extinction

GRADUALISM OR CATASTROPHISM?

What, precisely, is meant by the time scale involved in a mass extinction event? When geologists use the term, do they mean sudden and short-lived events on the order of days or decades or much longer periods of time? Is 10 million years too long for large-scale extinction to be considered one event? An extinction of numerous species that occurred in one day clearly seems to be an example of a catastrophic rate of change. But a million years looks much more like gradualism. Geologists have found it difficult to specify the duration of mass extinction events, and therefore they are unclear about the rates of change involved.

No matter how geologists answer these questions in the abstract, they must acknowledge that not even the marine fossil record is anything like a daily diary of life on earth. Because sedimentation rates can vary enormously, scientists have trouble determining how much time is represented by sediments in a particular environment. If scientists find the last of a particular species of mollusk in sedimentary rock located 3 inches below the imaginary line they have drawn at the end of the Permian, is the mollusk part of the Permian mass extinction? Geologists face two problems: They do not know exactly how much time is represented by 3 vertical inches of the sedimentary record, and apart from that, they can assume that not every species will be cooperative enough to preserve a fossil right at the time horizon they are trying to investigate.

This latter problem becomes even greater for large animals, which are more scarce in the fossil record because they do not densely populate any habitat, and also for land-based animals, which are only infrequently preserved. Compared with mollusks, dinosaurs contribute little to the fossil record: If geologists find the last dinosaur fossil 3 feet below the stratigraphic end of the Cretaceous, can they assume that the species in question disappeared in the mass extinction that marks the end of that period?

Here is another twist on the difficulties geologists face concerning the rate of change: Do some species go "extinct" simply because they gradually evolve into a new species? This kind of extinction surely is different from a mass dying on a sudden timescale. Indeed, it is only **pseudoextinction,** since much of the genetic material of the species lives on in what it has spawned. No less an authority than Darwin thought that extinctions resulted from gradual, not catastrophic, change, and he believed that most extinctions were only pseudoextinctions, since another species evolved in place of the old. As we have seen, Darwin knew that the generation of new species was one of the difficulties for his theory, but he was sure that speciation, and the resultant pseudoextinctions, occurred gradually. But as we have pointed out, modern evolutionary biologists see more scope for rapid change than their predecessors did. This is complemented by a de-emphasis on the pseudoextinction concept and the possibility of real and mass extinction events occurring on geologically short timescales. In sum, the new evolutionary theorists are much more inclined to entertain both sudden events of speciation and sudden extinctions.

THE CAUSES OF EXTINCTIONS

When geologists accept the idea of mass extinctions occurring in a relatively short interval of time, they are naturally led to speculate about the causes of such events. Unfortunately, the cause of extinction is generally not preserved in the fossil record, like a knife at the scene of a murder. Geologists have therefore postulated many different causes for particular episodes of extinction.

If the Big Five are meaningful events, special and distinct from normal, background extinction, and if they occurred over truly short time intervals, they seem to call for special explanations. The most famous mass extinction event is that of the Cretaceous–Tertiary (K–T) transition. (It is called the **K–T boundary** because, in German, the word for Cretaceous starts with a "K," and this helps limit confusion with such periods as the Cambrian and the Carboniferous.) The popular press has stated, over and over again, that the mass extinction at the end of the Cretaceous eliminated dinosaurs from the face of the earth. And in the 1980s, the K–T extinction gave rise to a theory about the impacts of meteorites that interested the press because it named a highly catastrophic global change as the mechanism for mass extinction.

THE K–T BOUNDARY

Geologists have long known that many animal species on land and in water that prospered in the latest Cretaceous period—a time known as the Maastrichtian—did not survive into the Tertiary. In the last century, when geology was firmly in the hands of gradualists, the apparent abrupt and massive extinction at the K–T boundary was explained away as a matter of principle. Methodology and assumptions, remember, underlie all scientific work, and the idea of gradualism permeated the best geologic research of the nineteenth and early twentieth centuries.

The grand dean of gradualism, Lyell, thought that the creatures present in the latest Cretaceous, that is, in the Maastrichtian sedimentary beds, simply could not have disappeared overnight. Such an event was contrary to the framework on which he had built his understanding of earth processes. But the fossil record in sediments seemed to show the abrupt departure of a number of species, followed in the record by a new and quite different set of animals. Lyell therefore postulated that for some reason, there had been little or no deposition at the end of the Maastrichtian and well into Tertiary times. As it happened, he was partially correct, for the K–T interval was a period with low-standing oceans and hence little deposition in many areas above sea level. But Lyell assumed that animals gradually speciated and became extinct. The rock record "must," therefore, have had a large gap in it. On this point, Lyell's assumptions about gradualism in earth history won out over the substance of the fossil record.

Recent research by paleontologists has created a fairly large descriptive database regarding the demise of many species at the K–T boundary. The evidence suggest the following:

> At sea: Bony marine fishes suffered heavy extinction at the K–T boundary; sponges, snails, clams and sea urchins lost many species and genera; ammonites went extinct but may have been in sharp decline during the Maastrichtian (that is, before the boundary); and with the exception of turtles, marine reptiles went extinct at the end of the Cretaceous, as did many—but hardly most—microscopic marine organisms.
>
> On land: Bony freshwater fish seem to have survived the boundary largely undamaged; amphibian species survived well; turtles lost some, but not most, species; crocodilian species survived intact; lizards crossed the boundary without incident; dinosaurs declined in the Maastrichtian and disappeared about the time of the K–T boundary; plants show little change, although they lost species a couple of hundred thousand years before the boundary; birds are too poorly preserved for us to know whether the K–T boundary affected them; marsupials lost many species and nearly followed the dinosaurs into nonexistence; and finally, a number of small mammals became extinct at or near the K–T boundary.

As you can see, the record is complex, and it grows more so as paleontologists discover more fossils. The popular press has often misrepresented the K–T boundary, implying that there is direct evidence that the dinosaurs were flourishing until a meteorite impact dispatched them into oblivion. The picture is painted of large dinosaurs lumbering in lizardlike ignorance to their doom. Mammals, the press has lectured the public, survived the disaster because they were warm-blooded or because they were small and could burrow down to wait out the disastrous consequences of the meteorite's impact.

Actually, as we have seen, many mammalian species did go extinct at or near the K–T boundary. And geologists believe that many dinosaur species were warm-blooded. Finally, according to some researchers, the dinosaurs were declining throughout the Maastrichtian. In short, it would appear that the K–T extinction, as it affected terrestrial species, was not related to body size or to warm-blooded versus cold-blooded status. We need to back up and take a look at the broader evidence before we can attempt to evaluate the famous meteorite theory.

A NUMBER OF HYPOTHESES

Scientific hypotheses concerning the cause of the mass extinction at the end of the Cretaceous have proliferated. The more significant include

1. A change in sea level with resultant climate changes.

2. An increase in volcanic activity over many years, the ash and gases from which caused a "volcanic winter" because the sun's light was screened from the earth.

3. The impact of a large meteorite on the earth, triggering a similar "winter" but one beginning abruptly on a particular day and hour.

The first point to consider about these three hypotheses is that some of the proposed changes are gradualistic, others catastrophic. Sea level, for example, could not drop 300 meters overnight. At the other end of the continuum, the impact hypothesis is extremely abrupt. One might hope that the rock record could tell us whether gradualism or catastrophism is the useful framework for the K–T boundary. Alas, the fossil record and the sedimentary rocks laid down near the end of the Cretaceous period give ambiguous evidence about the timescale of change.

Researchers who prefer to emphasize gradualistic change point to the fact that some marine animals, like ammonoids, declined through a million years or more and disappeared before the Cretaceous period came to a close. Similarly, most dinosaurs species died out before the K–T boundary; some researchers argue that the remaining dinosaurs seemed to have been in decline during the Maastrichtian. But those scientists committed to catastrophism like to note that many species of marine plankton, for which we have a good fossil record, died out abruptly just at the K–T boundary.

Setting aside fossils, what can the rocks themselves tell us about the nature and rate of change at the time in question? Stratigraphers who have studied the late Cretaceous believe there was a regression (lowering) of the seas several million years before the boundary with the Tertiary. When the Cretaceous period ended, it now appears, a modest transgression (rise in sea level) was occurring. Stratigraphers believe that the last 8 million years of the Cretaceous were abnormally cold. Gradualism, it would appear, can account for much of what we see in the rock record. But it cannot explain everything.

THE BOUNDARY CLAY

A thin (1 cm) layer of marine sediment occurs widely around the world, just at the K–T boundary. It is distinct and clear in many localities, and it may be no surprise that researchers fond of catastrophism focus their thinking on this layer. The layer of thin clay is abrupt enough to serve as the clear stratigraphic definition of the change from Cretaceous to Tertiary. In 1980, a famous physicist named Louis Alvarez and his son Walter, an accomplished geologist, showed that the clay layer contained an unusually high amount of an element called *iridium*. Because much has been made of this **iridium anomaly,** we need to consider the geochemistry of this unusual element.

Iridium is a metal that can be added to the surface of the earth from three sources. Cosmic dust, which is always falling to our planet from space—albeit at a slow rate—contains small but measurable amounts of iridium. The eruption of volcanoes is another source for iridium in sediments. Iridium is present in greater concentration in certain magmas, namely, mafic and ultramafic magmas which come from deep within the earth. Major volcanic eruptions from such sources can spew iridium-enriched particles into the upper atmosphere. Such particles are spread around the planet, as we ourselves have seen in the recent eruptions of Mt. St. Helens and Mt. Pinatubo. Last, iridium could be enriched in sedimentary clays because of the impact of a meteorite. Meteorites contain significant amounts of iridium, and if a large meteorite reached the surface of the earth, much of it would be vaporized. Particles of the meteor would be spread around the globe in a manner similar to volcanic ash.

To sum up the strength of the impact theory, we should note that we have every reason to expect that large meteorites have hit our planet during its long history. The heavily cratered surface of the moon indicates that our natural satellite has been struck many times, and we have no reason to think the earth is any different. Craters on our planet simply are harder to see, owing to more active erosion and to tectonic movements that "erase" features from the earth's surface. Thus we must assume that a number of substantial meteorites struck the earth in the past. As one noted geologist put it: "From everything we can gather, [meteorite impacts] must have had a dramatic impact on climate. If we did not already have evidence of biotic crises, this point would demand that we search for them."[4]

The iridium anomaly is undoubtedly important. Other lines of evidence favoring the meteorite hypothesis, or at least consistent with it, include:

1. The ratio of other unusual metals to iridium in the boundary clay is similar to the same ratios found in one kind of meteorite.

2. A layer with an iridium anomaly has also been found in nonmarine K–T boundary sediments in Montana, suggesting that whatever produced the anomaly was truly a global phenomenon affecting both land and sea.

3. Microscopic, "fluffy" carbon is present in the boundary clay and has been interpreted as soot resulting from worldwide forest fires triggered by the meteorite impact.

4. Grains of "shocked" quartz have been found in the boundary layer and are interpreted as resulting from the shock waves of an impact recorded in the grains at enormous pressures.

5. An impact site of the right age has been proposed in Mexico. It is a very large crater, extending into the Gulf of Mexico.

6. A meteor impact is best linked to mass extinction by looking at the marine fossil record or microorganisms, since that is the most complete source of fossils available. And when geologists study marine protists, they often see, although not always, that at the Cretaceous–Tertiary boundary, the world's oceanic ecology underwent a geologically rapid transformation just at or near the boundary clay.

JUST RECENTLY

The popular press has recently presented a modification of the theory it finds so newsworthy. Some scientists now believe that a large meteorite hit the earth at the K–T boundary and that its impact sent massive seismic waves throughout the planet. The waves would have been focused by the shape of the globe and could have triggered volcanism at the area directly opposite the earth from the impact site (the antipode). Thus, the volcanic activity and the shower of debris from the impact would have occurred together. The volcanism would have extended the time interval in which the atmosphere was choked by ash and, perhaps, sulfur compounds.

[4]In a letter (December 1994), Alfred Fischer gently reprimanded me for my doubts about the impact theory. Professor Fischer, a paleontologist and member of the National Academy of Sciences, has no such doubts, although he agrees that impacts cannot explain everything we see in the fossil record.

OTHER VIEWPOINTS

The lines of evidence in favor of a meteorite impact are interesting, and most American physical scientists generally consider them to be persuasive. Because the meteorite hypothesis partly originated through the work of an American physicist, we shouldn't be surprised that it has had a favorable reception in that community. American geologists are somewhat more balanced in their views about the meteorite hypothesis. European geologists and paleontologists have been the least impressed by the iridium anomaly and all that has flowed from it.

Evidence against meteorite impact, or against the importance of any impact to the extinction record, includes:

1. There is stratigraphic and paleontological evidence that many of the extinctions at the end of the Cretaceous period were not simultaneous and that marine and terrestrial extinctions did not occur at exactly the same time.

2. Freshwater reptiles are sensitive to temperature, yet they survived the K–T boundary almost untouched, as did other species in a manner not consistent with the impact hypothesis. One must infer that the cause of the extinction was much more complex than physicists envision.

3. Fluffy carbon (soot) is present at many points in the stratigraphic record, not just parts of the K–T boundary. It results, geologists believe, from local fires.

4. The demise of the dinosaurs, which is the part of the meteorite impact theory that has the most appeal to the public, may have occurred before the K–T boundary. There is a 2-meter gap between the last dinosaur fossil of the Maastrichtian and the iridium anomaly. In fact, all questions about the meteorite and dinosaurs are poorly constructed, for stratigraphic control on the demise of the dinosaurs is necessarily fuzzy: Dinosaurs did not live in the environments best suited to fossil preservation, and there were not many of them (compared, say, with plankton in the ocean). Geologists simply do not know precisely when the last dinosaur roamed the earth, but some researchers think they do have evidence that many dinosaur species were declining during the Maastrichtian.

It is fair to say that the K–T boundary will be a source of continued debate for years as geologists and other scientists struggle to determine rates of sedimentation, the speed with which certain species met their demise, and the fundamental character of mass extinction events.

Human-Caused Extinctions

EXTINCTIONS AND *HOMO SAPIENS SAPIENS*

Let us close this chapter with a brief mention of modern extinctions and our own species. We humans have knowingly driven some species, such as the dodo bird, into extinction. Scientists also think that our ancestors in the Pleistocene may have been such successful hunters that they unwittingly forced some of their prey into extinction. Today, as we modern humans change much of the habitat of the globe to suit our agricultural techniques, we are driving even more species into extinction. The loss of rain forests, old-growth timberlands, and natural prairie is taking its toll on thousands of species.

Is *Homo sapiens sapiens* creating a sixth entry on the list of mass extinctions? Our rapid destruction of many natural habitats has alarmed some scientists. Here are just two short examples of this concern:

> The Earth's biota now appears to be entering an era of extinctions that may rival or surpass in scale that which occurred at the end of the Cretaceous [that is, the K–T boundary]. As far as is known, for the first time in geologic history a major extinction [will be caused] by a single species—*Homo sapiens*.[5]

> It is hypocritical to blame peoples of emerging, especially tropical, nations for following our own destructive paths of habitat alteration. It is self-delusory not to acknowledge that it is we of the technologically advanced societies who are buying South American beef for our fast-food hamburgers and wood clear-cut from tropical forest stands.[6]

Such scientists feel that we must, for our own sake, slow the rate at which we are destroying natural habitat. A diverse global ecosystem, enriched with

[5]Paul R. Ehrlich, "Extinction: What Is Happening Now and What Needs to Be Done," in *Dynamics of Extinctions*, ed. David K. Elliot (New York: Wiley, 1986), p. 157.

[6]Niles Eldredge, 1991, *The Miner's Canary: Unraveling the Mysteries of Extinction* (Englewood Cliffs, NJ: Prentice-Hall, 1991), p. 218.

as many species as possible, is essential to the planet's biological stability. Therefore it also is important to our own survival.

Although no one would dispute that a complex ecosystem has positive value for us and for the planet, the human cost of changing our ways is staggering. Most of the world's peoples are desperately poor, and restricting further changes in habitat would increase hunger—and death—in many countries. So far there has not been a useful public debate, much less an international consensus, about how we want to shape the earth's habitats and our own, as well as our fellow species', existence.

Should Scientists Comment on Human-Caused Extinctions?

Scientists are always wrong to think they are above personal and cultural values, but they can usefully contribute technical knowledge to the discussion of habitat and extinction. It is, indeed, important for the public to understand that we seem to be well on our way to causing as much death and destruction as any meteorite may ever have done.

But it also is important for the public to keep in mind that scientists research and address only particular types of questions, such as problems in the empirical realm of efficient causality. Scientists are not experts about public policy, cultural values, or the spirit of the age. Besides, the considerations regarding our biosphere and extinction rates are so complex that scientific specialists do not yet have the expertise needed to understand the systems and predict cause and effect.

Summary and Conclusions

Extinction is a recurrent part of the fossil record: The overwhelming majority of species that have flourished through earth history are now extinct. Geologists often note that our presence in the world seems to be contingent on a long series of historical accidents. Whether the natural order is contingent, however, is a separate question.

There are five periods of higher-than-normal extinction rates that many geologists believe demand separate explanations. Researchers who emphasize the possibility of catastrophic change see evidence that mass extinctions are sudden and short and that at least one of them was triggered by a meteorite impact. Although it presumably is true that meteorites have struck the earth at times throughout its history, many paleontologists are not convinced that an impact caused all the changes in the fossil record of Maastrichtian times.

In any event, the methodological debate about catastrophism and gradualism clearly has changed; now all geologists accept the possibility of sudden and dramatic changes in earth processes. Scientists have come a long way from Lyell's authoritative insistence on gradualism.

Finally, as we have seen, our own species is triggering extinctions on a substantial scale. Both scientists and nonscientists are legitimately concerned about the thoughtless way we are altering our planet, which is, after all, our only home.

QUESTIONS FOR DEBATE AND REVIEW

1. Defend the theory that a meteorite impact contributed to extinctions at the K–T boundary.
2. Defend the idea that mass extinctions are illusory; that is, they are not special types of events and therefore do not need separate explanations.
3. Dispute the theory that a meteorite impact contributed to extinctions at the end of the Mesozoic era.
4. Describe the anthropic principle.
5. What difficulties do we face when trying to prove the importance of contingency in the natural world?

CHAPTER 12

"Creation Science" Versus the Historical Sciences: The Debate and the Law

The age of the earth, the origin of life, and the relationship of humans to other animals are topics that interest both geologists and others in our society, including the religiously devout. Much energy has been spent in the United States debating what children in public schools should be taught about these matters. The argument is often deeply emotional, and it always has political and legal dimensions. Perhaps it is not surprising that the debate tends to wander away from strictly scientific, or theological, considerations. Certainly, all too often, the main participants in the clash have descended to simple polemics. It is therefore highly difficult to achieve what our society needs: a carefully reasoned argument resting on clearly defined terms. Such a discussion should give all participants a legitimate place to stand—if only for the purpose of recognizing America's long-standing pluralism, or diversity, and thereby enhancing potential intellectual arguments to be considered in the future.

We will consider our society's divisive debate about these matters because it brings us to the heart of what scientific work is and how it affects other areas of our lives. We must begin by identifying the participants, most of whom believe (often correctly) that the other groups have misrepresented them.

The different participants in the current clash might be roughly grouped together as follows:

1. Biblical literalists like the "creationists." "Creation scientists" believe, as a matter of religious conviction, that the earth was created by direct action from God in one week about 4000 B.C.E. Humans were created specially by God and with a spirit that all other animals lack. Sedimentary rocks and the fossil record were formed, according to this view, during the flood of Noah's day. Biblical literalists represent one expression of the catastrophist school of thought. They hold that insights from empirical work are necessarily secondary to the absolute knowledge given to us in the revelations of scripture, usually thought to be best represented by the authorized King James translation of the Bible.

2. The Vatican's spokesmen, and most Protestant and Jewish theological authorities. These people mesh a theistic view of creation with some form of the scientific theories about the age of the earth and the cause of the varieties of life. They usually accept the framework of gradualism whenever the scientific community uses it. But they often think that the fundamental origin of early life remains in the hands of a divine being, and they are often heartened by scientists' inability to create, in the Miller–Urey type of experiments, organic molecules more complex than amino acids. This group is divided on what special qualities *Homo sapiens* may possess compared with those of other animals. Increasingly they accept the notion that humans are on some kind of continuum with other intelligent mammals such as primates, dolphins, and elephants.

3. Secular, scientific critics of Darwinism. This is a varied group of critics, both professional scientists (mostly in physics) and wide-ranging intellectuals, who are impressed by some of the difficulties in the Darwinian tradition. They thrive on the shortcomings of neo-Darwinism and the problems of the punctuated equilibrium viewpoint. They do not criticize the geologically accepted age of the earth, and they can be agnostic about the origin of life. Some members of this group accept the possibility of final causes operating in the universe and the biosphere. They believe that human beings are more similar to other conscious animals than they are different. Like Group 2, these people are often misunderstood by other participants in the debate, and they are often incorrectly lumped with religious extremists, on the one hand, or with members of the scientific establishment, on the other.

4. Defenders of neo-Darwinism and punctuated equilibrium. This group includes almost all paleontologists and evolutionary biologists. The neo-Darwinists

generally maintain a gradualist outlook, whereas the others allow for more rapid evolutionary change, but change that is still exclusively based on efficient causality. The most important vehicle for evolution is thought to be natural selection acting on variations within the population. Some members of this group view mass extinctions as catastrophic, but others still insist that most so-called mass extinctions are merely variations of background extinctions. All increasingly tend to think that extinctions are random, not directional.

A Brief History of the Controversy

At the beginning of this century, some American citizens organized themselves against the teaching of Darwin's theory of natural selection, a move reflecting our society's changing social and educational traditions.

For the first time, children were required to attend school well into their teens. Because many parents could not afford private schools for their children, they were thus forced to accept public education or break the law. Accordingly, this period marked the first time that virtually all teenagers in the country were exposed to at least some science class work. It was an adjustment for all concerned. When high school biology teachers taught the basic outlines of Darwin's theory, parents were sometimes shocked to hear from their children that God had little, if anything, to do with life on our planet. Furthermore, it distressed many parents that their children were being taught that humans evolved from the lower animals, as such views appeared to be atheistic. Concerned parents petitioned their school boards and state legislatures to eliminate the teaching of Darwin's theory, but enforcing laws against teachers who continued to expose their students to a respected scientific theory was difficult.

In 1925 the famous Scopes "monkey trial" attempted to enforce a Tennessee statute against the teaching of evolution.[1] Although John T. Scopes, a public high school biology teacher, was convicted of breaking the law, the trial became an important victory for the proponents of Darwin's theory. The publicity surrounding the trial portrayed the defense sympathetically:

[1]A delightful and highly readable account of the trial is given by Ray Ginger in *Six Days or Forever?* (New York: Signet Books, 1960).

Scopes and his defenders were presented as informed, enlightened, and unbiased, and opponents to Darwinism were made to look ignorant and hateful. After 1925, although anti-Darwinian statutes were not introduced to state legislatures, many biology textbooks, in order to avoid controversy, nevertheless eliminated all discussions of life's history from their pages.

For decades, Darwinism was seldom taught, even though it was not legally prohibited. But in response to the Soviet Union's launching of *Sputnik* in the late 1950s, the United States Congress began pouring money into science education at all levels. Science curricula were drawn up as models of what the public schools should teach their students, and Darwinism was included as the foundation of high school biology. In a few years, textbooks reflected these government recommendations, and so evolutionary theory appeared, once again, in the public school classroom.

By this time, the scientific theory had become much more sophisticated than it was in Scopes's day. But opponents of the theory still objected to Darwinism in any form, believing it to be an atheistic and false view of the origin and evolution of life. Slowly, scientists and intellectuals began to understand that the noise was not going to die down, that there was an important social and political problem bound up with their work.

Many biblical literalists, who call themselves **"creationists"** or **"creation scientists,"**[2] began to try to restrict what biology teachers could teach in the public schools. Once again, state legislatures passed statutes limiting the teaching of Darwinian theory or requiring that biology teachers give "creation science" equal time in their lectures and assignments.

By the 1970s, animosity on all sides of the debate was firmly cemented, and social and religious conservatives offered anti-Darwinist laws to many state legislatures. During this time, biblical literalists had become much more sophisticated and so began to argue that Darwinism in all its forms functioned as a religion because it tried to explain the origin and meaning of life. This "religious" view, they sometimes contended, should not be forced on their children. On the other hand and at the same time, biblical literalists began calling themselves "creation scientists" and maintained that all they wanted was to present the scientific shortcomings of Darwin's theory or to offer other scientific theories about the origin of life and the nature of the fossil record.

[2]The terminology here is confusing because the Vatican, mainline Protestant authorities, and Jewish and Muslim theologians also are creationists, that is, believers in a created world. Biblical literalists like to claim they are Christianity's only "creationists," but this is simply not the case. To make matters more confusing, most scientists object to the term *science* as biblical literalists apply it to themselves.

BOTH SCIENTISTS AND "CREATION SCIENTISTS" HAVE BEEN SLOPPY

As the "creationists" take delight in noting, scientists have stated, on many occasions, that Darwinian evolution is the only possible explanation for our existence. Scientists have even asserted that Darwinism explains the fabric of morality and the "meaning" of life. Indeed, an authority on neo-Darwinism published a book entitled *The Meaning of Evolution*[3] that ends with an exploration of morality's biological foundations. Members of our society who are in no way part of the fundamentalist right wing can and do take offense at such discussions in a scientific work. Science, many casually assume, deals with facts and should not try to teach values. Therefore, scientific research cannot instruct us about the "meaning" or the "purpose" of life or about morality.

On the other hand, biblical literalists are not proposing scientific alternatives to Darwinian thought. Rather, they are motivated by what they take to be God's personal and spiritual call; they do not accept any empirical inquiries into the natural world except insofar as such work can confirm a literal reading of Judeo-Christian Scripture. What they propose is not "science," and their use of the term is merely a skillful rhetorical device to obscure the fundamental differences between biblical literalism and other approaches to knowledge.

THE HEART OF THE DISAGREEMENT

The debate has become confusing in the public's mind because all sides employ abundant "us-versus-them" rhetoric. There are, however, more than two viewpoints at issue, an important facet of the problem that has been overlooked or obscured by politicians and the courts alike. Let us summarize the several different methodological and empirical issues.

Darwinian Evolution. Gradual and naturalistic, this classic theory rests on the efficient causality of natural selection. It points to change within a species (microevolution) and argues for change from one species to another. The theory accords poorly with the fossil record, and empirical evidence regarding speciation events is still regrettably sparse. But the theory is powerful and useful because it explains taxonomy and many other elements of biology.

[3]George Gaylord Simpson, *The Meaning of Evolution* (New Haven, CT: Yale University Press, 1949).

Neo-Darwinian Evolution. Also gradual and naturalistic, this variation of the classic theory also rests on the efficient causality of natural selection. But selection, proponents argue, can occur at the level of the kinship group or the gene as well as the individual. This view argues for change within a species (microevolution) and from one species to another. The theory accords poorly with the fossil record for the same reasons as Darwinian evolution does. Again, the empirical evidence regarding speciation events is frustratingly sparse. But the theory is powerful because it explains taxonomy and many other elements of biology.

Evolution via Punctuated Equilibrium. This approach is less gradualistic but still entirely naturalistic. It rests completely on the efficient causality of natural selection. It, too, argues for change within a species (microevolution) and from one species to another. This theory fits the basic characteristics of the fossil record better, just as it was designed to do. Once again, the empirical evidence regarding speciation events is regrettably sparse, and yet this theory relies heavily on speciation. But the theory is powerful because it explains taxonomy and many other elements of biology.

Other Forms of Evolutionary Theory. These theories are even less gradualistic, and they may or may not be naturalistic.

View 1. Evolutionary nihilism. This position is similar to that of neo-Darwinism and punctuated equilibrium but emphasizes the importance of the random processes of mass extinction. No kind of natural selection or overriding drive is as important to a species as "evolutionary luck" is. This theory probably fits the fossil record best, but some mass extinction evidence is still highly debatable.

View 2. Secular final causes. This is actually a group of theories positing a drive toward increasing complexity, self-consciousness, self-organization, or the occupation of all biological niches. These theories rest on final causality until and unless we can discern efficient causes for what we see. All such theories allow change within a species (microevolution) as Darwin sketched it. But Darwinian selection is not the cause of change from one species to another. Such theories can fit the fossil record well, but so far they do not empirically explain evolutionary mechanisms or the nature of the "drives" proposed for the natural world.

View 3. Theistic evolution. This approach proposes a framework for recurring episodes of divinely guided creation of the increasingly complex life-forms we see in the fossil record. It rests on final causality and is not naturalistic. It allows change within a species (microevolution) as Darwin described. But the change of one species to another is caused by divine, not natural, action. The theory fits the fossil record well but, by definition, is not within the realm of empirical testability. The enormous depth of geologic time and the abundant evidence of extinction in earth history appear to make the Creator less than highly efficient. This theory offers no explanation, except divine will, for the regularities of taxonomy. Biological theory before Darwin was generally similar to what theistic evolutionists defend today.

Biblical Literalism. This view states that instantaneous and nonnaturalistic creation occurred about 6000 years ago in the span of one week. The theory allows change only within a species (microevolution). Any change from one species to another, the theory insists, is illusory. This viewpoint fits the fossil record very poorly, and it is not within the realm of empirical testability. It offers no explanation, except divine will, of the regularities of taxonomy. *It also does not explain why other religions have a different understanding of the world's creation.*

There are many interrelated questions that the "creation scientists" have brought before our society. Unfortunately, the questions require careful thinking and reasoned answers, qualities that public debates often lack. These questions include

- Where does science—especially of the sort now being practiced by Francis Crick—end, and where does religion begin? This then leads to the questions, Whose religion is important in our society? Must we value the claims of every Protestant splinter group?
- What should the state require teenagers to learn about the origin and evolution of life as described by current scientific theories?
- What, precisely, is evolutionary theory? Does it address the ultimate cause of life's earliest expressions on this planet?
- Does all evolutionary theory conflict with religious values and beliefs as expressed in Western society through the Jewish/Christian/Islamic tradition?
- How can we broaden this discussion to include the millions of Americans of non-Western extraction whose religious beliefs differ from those expressed in the Judeo-Christian Bible?

THE CONTINUING DEBATE

The most recent legal case about these matters was decided by the United States Supreme Court in response to a 1981 Louisiana law requiring a balanced treatment of "evolutionary science" and "creation science." On the one hand, the biblical literalists argued before the court, they wanted only a chance to present scientific evidence against Darwinian evolution. On the other hand, they asserted that there was no real difference between the creationist and evolutionist positions, since both functioned as a religion.

In 1987 the Supreme Court voted seven to two to strike down the Louisiana law as a violation of the First Amendment's establishment of religion clause. The court maintained that science and religion were meaningfully different and that Darwinian theory was science, whereas biblical literalism was a religious tenet, one segment of Protestant fundamentalism. In our diverse society, the Constitution prohibits the teaching of one religion in the public schools. Since biblical literalism is only one religion among many, teachers cannot present it by itself to public school children. And since biblical literalism is not science, it cannot be taught by biology teachers in the public schools. Scientists, most religious leaders, and theologians throughout the country were pleased by the decision.

Justice Anthony Scalia wrote the dissenting opinion in the Louisiana case, resting his argument on the idea that scientists use an evolutionary framework of thought to explain not only changes in species but also the origin of life itself, a subject that Scalia regards as religious. Indeed, as we have seen, the chemical or prebiological evolution of organic molecules is often discussed by scientists in their research work on life's origins. Scalia reasoned that creationism also addresses the origin of life and so could be taught as an alternative to the Darwinian framework.

Professor Stephen Jay Gould, responding to Scalia's opinion in *Natural History* magazine, denied that evolutionary theory applies to the origin of life. Gould stated that in Darwin's words, evolution addresses only "descent with modification" from prior living things. Although Gould believes that the origin of life is a proper problem for science to address, he insists that evolutionary science does not include a theory of life's origins.

But within the framework of Darwinian selection, many evolutionary biologists do speculate about the origin of life in the early earth's oceans. Evolutionary biologists have not always limited their thinking to "descent with modification" from previously living organisms. At least when doing this kind of work, they are speculating about matters that used to be left to myth or theology, not science. Although it is not surprising that some parents are offended by such speculation, it could serve as a useful starting point of a discussion of where the edge of science lies and what constitutes the spirit of empirical inquiry.

In addition, just as the "creation scientists" contend, prestigious supporters of Darwinian theory have written about morality and the meaning of life in evolutionary terms. No less an authority than George Gaylord Simpson, for example, concludes his book, entitled *The Meaning of Evolution*, with a section on evolutionary ethics. More recently, Harvard biology professor E. O. Wilson and his followers established a whole new subdiscipline within evolutionary theory. Their writings discuss morality in terms of *sociobiology*, a fully materialistic framework of thought resting on Darwinian natural selection. Again, evolutionary scientists are encroaching on what we, as a society, have traditionally recognized as religious turf. We might note, of course, that since Galileo's time, science has been pushing back the edges of the religious realm. Perhaps this case is no different, and we might argue that it would be detrimental to science to step back at this critical juncture.

Nevertheless, many members of our society who have no sympathy for "creation science" might still ask, If evolutionary theory leads, over and over again, to pronouncements about the origin of life and theories about morality, shouldn't traditional religions be given equal time somewhere in the school curriculum? Perhaps in response to this view, scientists could set aside their thinking about the origin of life and its meaning and about the moral or ethical implications of evolutionary theory. Surely at present, our diverse and pluralistic society has no consensus about what science can or should teach us about such matters. We all could embrace the view that the world is much richer and deeper than anything yet defined by empirical and nonempirical disciplines of thought. Therefore, understanding conflicting scientific theories actually magnifies the glory of what we *Homo sapiens sapiens* have accomplished in the realm of abstract thought.

In this regard, scientists should not be hesitant to say that their work is based on theory.[4] Recall that unlike the "theories" put forward in colloquial speech, scientific theories are a synthesis of measurement, calculation, experiment, and critical thinking. In scientific work, even though theories rigorously organize a great deal of empirical evidence regarding the natural world, they are not static and complete truths. For example, plate tectonics is still a theory, and geologists are able to discuss its shortcomings without the defensive fear

[4]Popular representations of research in evolutional science today all too often claim far too much for the scientific work on which they report. See, for example, Jonathan Weiner, *The Beak of the Finch* (New York: Vintage Books, 1994), which maintains that two Princeton biologists have seen Darwinian natural selection change the finches on one of the Galapagos Islands. Actually, the research has shown that the average size of the birds' beaks increased in drought years and decreased in wet years. No net change, however, has been documented, and the results would be regarded more accurately as variation within a population, not direct evidence of a new species coming into existence

or self-righteousness marking some of the legal testimony defending Darwinian evolution. Geologists adopted the plate tectonic framework because it explains so much about the world and draws on so many different types of data. It is, indeed, a good story! But the Columbia Plateau flood basalts in Washington State have not been well explained by tectonic theory, nor do we understand the mechanism that drives the plates' movement. Should geologists therefore view these difficulties as simply "unexplained at present" or as a fundamental threat to the whole tectonic framework?

A THOUGHT EXPERIMENT

Consider what might be the situation if our society had a somewhat different social structure. What if the biblical literalists had ignored Darwin and instead taken deep offense at plate tectonics? The scenario is not so far fetched: "Creation scientists," in fact, do not accept plate tectonics or the geological evidence for the earth's amazingly great age. (But so far, biblical literalists have gone to court about the teaching of Darwinian evolution rather than about the tectonic theory of the earth.) If fundamentalist clergy were preaching against the basic finding of tectonics, geophysicists and geochemists might well be defensive enough to lose sight of the nature of scientific work, just as paleontologists and biologists in the existing public debate sometimes do. But in the peace given to some geologists by the diversion of "creationist" energy toward combating Darwin, scientists can feel comfortable saying that plate tectonics is a theory. It is not immutable. Indeed, it has changed significantly since it was introduced, for example, by the inclusion of ideas about "suspect terrains" and their importance in the western United States and Canada. Tectonics delights scientists because it explains so much, but they could—and should—abandon the whole framework of tectonics if a better empirical theory surfaced. This is not a criticism of the theory; it is, in fact, a credit to the amazingly successful methodology at the core of all science.

Similarly, religious-minded members of our society could hold in respectful attention the idea that all empirical research and theological debates magnify the glory of any creator or first cause. In fact, that is the viewpoint of most church members in this society. The biblical literalists are only a very small minority of the Jewish/Christian/Islamic tradition. "Creation scientists" are out of step with the Vatican, with widely recognized Protestant and Jewish theologians, and with all non-Western religions. Perhaps we can look forward to the day in which members of the mainline churches and non-Western religions take an active stand against "creation science" rhetoric and legal actions.

What Is Happening Now?

Unfortunately, the "creation scientists" have adapted to their defeat in the courts by changing tactics. They now are focusing on altering the content of high school textbooks and have been remarkably effective in this task. Rather than run the risk of controversy, many biology texts now shy away from presenting Darwin's theory or discussing its merits and shortcomings. Information about the cell, about heredity, and about ecosystems is presented without any overall theory uniting them and without a discussion of the changes in the history of life that we see in the fossil record.

Taking basic scientific theory out of the biology classroom is not, in my opinion, fair to the students. High school biology students assume they are being taught the most current biological science. Although scientists won the legal battle in the courts, we all are now losing the struggle to teach the next generation the outlines of one of science's most intriguing debates. The point is not just academic. Our society is becoming increasingly affected by a whole host of issues related to evolution, for example, in the rapid adaptive changes in strains of pathogens to our antibiotic drugs and the similar changes of insects in response to insecticides for our crops.

Summary and Conclusions

All variations on Darwinist theory conflict with a literal reading of the book of Genesis. Although the same could be said about plate tectonics, some biblical literalists have focused on eliminating any theory of evolution from biology classes in public schools. In general, the courts and the general public have not been sympathetic to the biblical literalists. However, the debate continues because it rests on fundamental concepts such as what constitutes science and what constitutes religion. Regrettably, rather than engaging citizens in a discussion of these important issues, both scientists and biblical literalists often descend into polemics.

"Creation scientists" have turned to another tactic: They are now concentrating on becoming members of school boards and members of panels that select textbooks for public schools. As a result, many high school biology textbooks omit serious discussions of Darwinism in order to avoid confrontations with the well-organized "creationists." We can only hope that mainstream Christian and non-Western members of our society will join the fray and help balance the views of zealots in the "creation science" movement, on the one hand, and of defensive scientists, on the other.

QUESTIONS FOR DEBATE AND REVIEW

1. Were you taught some form of evolutionary theory in high school? If so, how does what you were taught fit into the range of evolutionary theories current today?
2. If you are a participant in organized religion, find out what your religious leadership says about the age of the earth and the evolution of life. Where in the spectrum of evolutionary debate do you find yourself?
3. What do you think should be taught about the origin and history of life in science classes in our public schools? Defend your views.
4. What do the non-Western religious traditions teach about life's origins and history? How do members of non-Western religious groups feel about "creation scientists"?

CHAPTER 13

Methodological Limitations Revisited

We have come a long way. We have seen that previous civilizations often relied on supernatural explanations for the processes we see on the earth around us. Following the ideas of ancient Greece, Westerners slowly began to look for scientific explanations for events in the physical world. For a time, geologists and paleontologists tried to match their work to the biblical account of creation in Genesis. But the rock record demanded a much longer history for the earth than anything the biblical authors had imagined, and the enormous complexity of the fossil record did not square with scripture.

Geologists then focused on what they did best: describing and interpreting hand samples from all over the continental crust. Ideas about rocks and the rock cycle, and about fossils and the depth of geologic time, blossomed. The biological theory of evolution and the geophysical theory of plate tectonics bring us up to present-day research in the earth sciences.

One of the points I have repeated over and over in this book is that scientific research is a human endeavor. Like all our intellectual labors, science is sparked by creativity, shaped by our culture, and marred by personal foibles. Contrary to the views of many members of our society, scientific research has deeply subjective components, including aesthetic values.

Geologists, it has been argued, are more like storytellers than robots. Earth processes are always embedded in time, and geologists try to reconstruct the story of earth history. But the tales told by scientists cannot be fantasy. Geological stories must fit the data, and they must undergo the restrictive and conservative process of peer review.

We also have seen that geologic research has progressed despite personal and idiosyncratic decisions about the very framework within which we examine our planet's history. After two centuries, the earth sciences are still marked by contradiction and confusion about uniformitarianism, the concept most often called geology's foundational principle.

The Earth Sciences

Contrary to many assumptions about the scientific method, geologists and paleontologists usually examine hypotheses that cannot be tested directly. Geologists cannot manipulate their most important variable: time. Researchers therefore seek circumstantial evidence in favor of a given view. Such evidence may have been gathered originally to address a much narrower question, of interest to only a few specialists. Data of this sort are not only challenging for an "outsider" to interpret, but they can even be difficult to locate in unfamiliar journals with a small circulation.

What happens when geologists encounter a new hypothesis? One might hope that if geologists can find abundant and varied data in its favor and if the underlying ideas have the appeal of simplicity, scientists will conclude in favor of the new hypothesis. The history of the theory of continental drift bears out such hopes in one instance: The theory was accepted because it fit well with data from a wide variety of subdisciplines in the earth sciences and because it seemed to simplify the scientific understanding of many earth processes. Contrary to what a layperson might assume, changes in overarching theories have made relatively little difference to ongoing research in the subfields of geology, although they do influence such research in the long term.

The Age of Science

This book has necessarily emphasized that scientific hypotheses are always considered in human, historical circumstances. As one of the least precise sciences, geology particularly can be shaped by general trends in intellectual history.

We live in an age that exalts scientific work and seeks the objectivity of fact. This gives science a unique, indeed revered, status in our culture. Scientists claim to adhere to a rigorous methodology, an assertion that further enhances their status with the public. Laudable though their methods may be, they sometimes take them to be a description of the world. This is always a grave error, for the physical world may or may not follow the principle of parsimony or satisfy scientific standards for "beauty." At least part of the conflict between science and religion, and between philosophies of fact and value, festers in the gap between science's methodology, on the one hand, and poorly considered substantive claims, on the other.

One indication of the importance of science in our culture is the government's funding for scientific research. Other than a brief mention of financial issues in the Introduction, we have not been able to explore the mechanisms and the magnitude of funding for the geological sciences today. You should know, however, that in the distant past the U.S. government did not fund any type of scientific research. Private industry and wealthy individuals were the source of whatever money scientists could raise. Next, modern industry supported research, but it then largely stepped back from that role. During the cold war, the federal government poured money into scientific research in a host of disciplines. But future funding for scientific research is difficult to divine. Congress may change what academic research it supports in light of political pressures to cut federal spending as a whole. The geological sciences will be, it seems, increasingly hard pressed. Indeed, one congressman has now put forward a plan not only to reduce the earth science budget of the National Science Foundation but also to eliminate the Geological Survey (a long-standing branch of the Interior Department).

This is not only a fiscal matter. We live in a time marked by an undercurrent of cynicism and a generalized suspicion of specialists and specialized knowledge, and the American habit of anti-intellectualism may be coming to the fore once again. Certainly some members of our society are deeply distrustful of scientists even while most of us acknowledge the successes that science and technology have brought to our economy and society.

Why Care About Science If One Isn't Involved in It?

The conflict between the scientific and nonscientific factions of our culture is important to what we, as a society, may hope for our future. If fact can be separated from value, as many scientists and most intellectuals assume, then science cannot help us with questions of value. This means, among other things, that science and technology will not, by themselves, enhance our welfare. We therefore should not look to scientists for guidance regarding political, economic, social, or moral questions. But we generally do exactly that, expecting science and technology to rescue us from global problems. Although illogical, such appeals are understandable given the prestigious place that science occupies in our culture.

But there is another possibility: If fact cannot be separated from value, as premodern and postrationalist theorists and a few scientists (such as some sociobiologists) maintain, then science is not especially factual in a way superior to other disciplines. This means that science does not merit special status in our culture. Such a view lessens the gap between scientific research and questions of value. The opinions of scientists about world peace, human equality, and moral values may properly be consulted as we distance ourselves from the distinction between fact and value.

Evolution and All That

Our society's confusion about methodological and substantive claims has surfaced, over and over again, in the debate about teaching evolution in the public schools. The rhetoric on all sides has often been designed to appeal to our fears rather than our intellects. Both fundamentalist Christians and indignant scientists often prefer to set up straw figures as opponents rather than to attend to the conflict's central issues. Darwinism, as we have seen, is a theory that originated in the last century. In regard to the fossil record, evolutionary theory now has two main variants, neo-Darwinism and punctuated equilibrium theory. Much remains to be resolved. To be honest, scientists must admit that their thinking about evolution may someday be complemented by another completely naturalistic explanation of the change in life-forms in addition to natural selection operating on random genetic changes.

But problems with different types of Darwinist theory are rarely discussed by scientists when they think the public might be listening. This is not only contrary to the scientific spirit, it is a debilitating, fearful habit. All too often scientists have publicly promoted a simple notion of Darwinism as "factual." It does seem that life has had a long and varied history. That is as much of a "fact" as geologists will ever find. And species seem to have descended, in some sense, from common ancestors. But scientists are still struggling to understand the mechanism at work in evolutionary changes from one species to another.

Nothing in the Genesis account, if taken literally, accords with the complex record of fossil evidence or with basic notions about the age of the earth and the successive and distinctive eons of geologic history. Furthermore, the broad teachings of the Judeo-Christian tradition concerning creation have not been conspicuous in making room for the many species that geologists know have come into existence over long periods of time. Nor has Western religion made any significant attempt to explain the phenomenal death rate of species throughout the earth's history. Nor do biblical literalists make room in their teachings for the scores of venerable creation stories other than the Genesis account.

Our Successes and Our Limitations

Despite scientists' occasional shortcomings, which this book has briefly sketched, it is impressive that scientific research has described and successfully related a wide variety of physical phenomena. At its best, science is predictive, certainly no small feat in a world governed by change! Scientific research is a demanding and creative endeavor that has delighted generations of bright minds and has explained seemingly incomprehensible puzzles. Furthermore, scientific work has excelled in spawning technologies more complex than anyone could have envisioned a few decades ago, technologies that have revolutionized our personal, medical, social, and economic lives.

Nevertheless, we are not warranted in blindly assuming that scientific methodology and the structure of mathematics on which it stands are fundamentally related to the material world. We often unconsciously adopt such a view because of all that science and technology have accomplished. But blurring the differences between scientific methodology and substantive claims about the physical world is contrary to the best, most rigorous spirit of scientific research.

Nor should we simply assume that the methods of science are the best route to all forms of knowledge. Perhaps we have not yet found methodologies that future *Homo sapiens sapiens* will use to good effect. Although it is true that science has been extremely successful in what it does—indeed, in many ways it is our most successful intellectual method—the inherent limitations of science should not be forgotten. The philosopher Thomas Nagel reminded us of this point:

> Scientism [the view that science, and science alone, leads to knowledge] . . . puts one type of human understanding in charge of the universe and what can be said about it. At its most myopic, it

assumes that everything there is must be understandable by the employment of scientific theories like the ones we have developed to date. . . . [T]oo much time is wasted because of the assumption that methods already in existence will solve problems for which they were not designed; too many hypotheses and systems of thought are based on the bizarre view that we, at this point in history, are in possession of the basic forms of understanding needed to comprehend absolutely everything.[1]

[1]Thomas Nagel, *The View from Nowhere* (New York: Oxford University Press, 1986), pp. 9–10.

APPENDIX A

Facts and Values

Most educated members of our society accept without serious thought the assumptions supporting the methodology of twentieth-century science (rationalism, secularism, and the emphasis on efficient causality in the natural world). Science has explained so many things and fostered such a dazzling array of technology that the public doesn't often question its professional assumptions. It is generally thought that scientists determine the facts of the natural world according to the best possible means and, moreover, that these facts are not influenced by any subjective value given to the natural order. As Stephen Jay Gould wrote: "Science [is] an enterprise dedicated to discovering and explaining the factual basis of the empirical world [whereas] religion is an examination of ethics and values."[1]

Many parts of general university culture agree with this view. Students often are explicitly taught that "value judgments" are subjective and therefore individualistic. They are thus counseled to form their own views, to decide for themselves about moral issues such as abortion or resistance to draft registration and about what values they want to guide their lives. In this way, value judgments become the logical equivalent of personal taste in food: "Some people choose to have mincemeat, and others prefer pumpkin pie; there's just no accounting for taste."

But students also are often taught that scientific debates concern objective "facts" and that such facts are free of values. Indeed, science professors often present this view to their students.[2]

The matter is not so simple, however, and this brings us to a serious intellectual debate. *Some of this century's most distinguished philosophers cannot accept the distinction between facts and values that most scientists so blithely assume.* For one thing, "facts" are not free of the perceptions, beliefs, and

[1]Stephen Jay Gould, "This View of Life," *Natural History*, March 1994, p. 18. The same assumptions are reflected in the writings of many geologists, including the philosophically sophisticated S. Von Bubnoff, in his *Fundamentals of Geology* (Edinburgh: Oliver and Boyd, 1963).

[2]Some scientists have concluded that facts and values cannot be separated, because human values rest on the facts of genetics determined by human evolutionary history. Sociobiology is the principal branch of such thought, although some traditional neo-Darwinists subscribe to this idea as well. Most scientists, however, agree with Stephen Gould's sharp division of fact and value.

theories of the person observing them. As an example, two geologists can look at the same outcrop but take very different observational notes because they subscribe to different theories about how a particular rock type forms. As we have seen, the "facts" of the Channeled Scablands as J Harlan Bretz saw them were indeed different from the "facts" that other established geologists observed when they visited the same area. Where, one might ask, did we get the notion that "facts" are free of the observer's theories and values?

The framework of this question was constructed by the social and intellectual developments flowing from the liberal tradition. Note, however, that the term "liberal" in this regard has nothing to do with current progressive politics. Indeed, liberalism describes many of the values of conservative political parties, as well as the "liberal arts" of university life. Therefore, to try to limit confusion, I will use the term "rationalism" to represent this broad and multifaceted tradition.

Rationalist social and political thought was developed by writers like John Locke, Thomas Jefferson, and John Stuart Mill. It emphasizes the freedom of the individual (rather than the needs of a tightly knit group), religious tolerance (rather than the suppression of rival religions), and empirical standards for evaluating facts about the world. In this view, facts are to be determined by scientific research and are never to be equated with values. According to rationalists, although religious leaders can help us understand and live up to our moral obligations, neither religion nor philosophy can tell us how the material world is put together.

One highly emotional argument for considering the assumptions of rationalism is currently sweeping through some of our public schools. The "creation scientists," who claim that a literal reading of scripture should dictate what is taught in biology classes, are opposed to the very foundation of rationalist thought. Instead, "creation science" teaches that plants and animals were created directly by God and thus are valuable because they are his handiwork. These biblical literalists also deny that the earth is billions of years old, as geologists emphatically claim. Most important, biblical literalists do not look to science for any verification of their view of the world; rather, they rest their arguments on the revelation of God recorded in scripture. In recent years biblical literalists have begun to describe their views as "creation science," to equate rationalism with secular humanism, and to call both a religion. But most observers would agree that this is a matter of strategy meant to catch their opponents off balance, not an accurate description of what is at issue.

The fact that so many Americans identify with "creationism" in its broadest outlines is, according to most scientists, either an indication of the failure of science education in this country or evidence of the fundamental irrationality of *Homo sapiens sapiens* (especially, scientists often imply, poorly educated or religious people). Actually, the matter is not that simple: The many shortcomings within the framework of rationalist thought nourish the continuing debates about the tradition's basic assumptions. These shortcomings, many people argue, leave plenty of room for contributions from religion and other nonrationalist parts of our cultural tradition.

Social and political rationalism, all would agree, has made little headway in offering any theory of values. Current critics like to point out that if values do not rest on facts, why do we so consistently feel that some actions are immoral? The idea of torturing a two-year-old child for sport, for example, makes us all quite ill.

Rationalist thinkers have taken contradictory approaches to questions of value and morality. John Rawls and Ronald Dworkin are prominent modern philosophers who have offered thoroughly secular but highly different foundations for moral values. From their perspective, such disagreement is not fatal, for rationalists are used to thinking that important moral debates may well have no resolution. The philosopher Alasdair McIntyre and the theologian Karl Rahner, following in the footsteps of such writers as Reinhold Niebuhr, use another approach, which can be described as "postliberal" or "postrationalist." They retreat from the assumptions about the differences between "facts" and "values." Notice that this immediately puts them outside the group of intellectuals most naturally respectful of scientific methodology. Rahner pointed out that as moral creatures, we are grounded in our experience that the way we are is not necessarily the way we should be. This gap, however, is a matter of fact, not a question of abstraction. It is not a scientific fact, Rahner agrees, but that only shows *there are other kinds of facts in the world, facts that science cannot investigate.* Morality, from this perspective, is not so flexible as rationalists and most scientists believe. That is, moral questions do have absolute answers, which is what gives morality its gut-wrenching force.

The postrationalists, like most American citizens, are comfortable with the idea that moral values lie outside individual preference and prejudice.[3] There really are such things as right and wrong, the argument goes, and we must discover which is which in the problematic areas of our lives. We cannot simply choose, as earlier philosophers and many scientists supposed, to invest facts with moral value.

[3]See the data referred to by a professor of law at Yale University, Stephen L. Carter, in *The Culture of Disbelief* (New York: Basic Books, 1993). Chapter 11 is a good summary of the issues.

One way of summing up different approaches to the fact/value issue is to examine the changing use of the verb *to believe* in the English language. In the days of King James I of England and in the translation of the Bible he sponsored in 1620, *to believe* meant "to choose," "to wish for," "to value," "to cherish," or "to love." The value of what one "believed" in was as evident to the believer as the thing itself. This notion of believing in God meant that a person valued, cherished, nourished, and sought spiritual life. The facts of searching for unity with God's will were not separated from the value of the search. There was, no doubt, more and less fruitful belief in the lives of various people. That is, there was a difference between believing well and believing badly.

Today, the verb *to believe* generally means something quite different. Sometimes it refers to opinion, and people even use the word *belief* in this sense to denote personal opinions that run contrary to fact, opinions that we particularly dislike and wish to dismiss. For example, a scientist might say, "Ronald and Nancy Reagan believed in astrological predictions! Can you imagine such nonsense?!"

This change in the meaning of *believe* also reflects a fundamental change in the structure of our society's thought. The difficulties in the discussion of "facts" versus "values" have long roots and will be with us as far as we can envision the survival of our society. "Creationists" or biblical literalists, like the Roman Catholic Church of Galileo's time, are prerationalist in their approach to the relationship between facts and values. But biblical literalists do have something in common with important current theorists and many ordinary members of our society. Despite centuries of advances in science, the assumptions behind scientific methodology continue to confine empirical work to one realm, that of efficient causation and public observation. This way of explaining the world *alienates many ordinary people in our society, both religious and nonreligious, and some sophisticated intellectuals as well.*

Scientists, perhaps especially geologists, all too often bury their heads in the sand when these matters come up for debate. But they do a disservice to their discipline and to their own intellectual lives when they refuse to discuss fundamental assumptions and the resultant nature of empirical research. In addition, they should remember that the basic framework of science relies on the Jewish tradition as well as on Aristotle. For example, it was Western religion that fully developed the sense of linear time (the idea of a beginning, middle, and end). Earlier civilizations and non-Western traditions relied on cyclical time (an endless repetition of alternating states or cycles). But scientific ideas like evolution can be understood only in a linear framework, that is, one resting on the view of time found in the Bible. And as Christianity developed, it also planted the idea in our minds that only truth leads to true freedom but that such truth is difficult to discover. Scientists have used all these basic notions to good effect in the scientific ethos. Nevertheless, most of them are almost deliberately unaware of these debts and frequently condemn religious tradition in the most casual, and ignorant, manner.

Religious critics of science like to capitalize on the philosophical ignorance and lethargy of many scientists. From a religious standpoint, of course, there is no difference between fact and value.[4] Moral and empirical truths are interwoven, and so God's will is not something we can evaluate to decide whether we want it to prevail. Rather, the fact of divine will *determines* our values. For many religious people, this point is obvious. But social rationalism and scientific work rest on another perspective and therefore generate different theories of knowledge.

These disputes are not just intellectual battles. Their resolution (or lack thereof) is important to our species' future. We cannot reasonably look to science for answers to the problems created by empirical research and industrial society if we do not understand what kinds of problems scientists can address. The matter is not just academic.

[4]Some scientists are actively involved in organized religious life, as I am. But for the purpose of discussion, we can separate the practitioners of science and the practitioners of religion.

APPENDIX B

TERMINOLOGY FROM LIFE'S HISTORY

The following organisms, important to the evolutionary history of life on our planet, are briefly described for the benefit of those readers who do not have a background in biology.

Brachiopods Double-valved marine animals that feed on suspended material. They first appeared in the Cambrian period and resemble bivalve mollusks. They still flourish today.

Bryozoans A group of colonial animals still existing in both salt and fresh water and known informally as "moss animals." Like corals, they are composed of individual polyps occupying the interstices of a rigid skeleton. They feed on microorganisms that pass by in the water. Bryozoans became important on the seafloor during the Ordovician period.

Cephalopods Marine predators with tentacles for holding prey and hard beaks for picking apart small prey. Squids, octopuses, and their shell-bearing relatives are examples still flourishing today.

Conodonts The term *conodont* can be confusing. People occasionally use the word to refer to a Paleozoic, eel-like animal with teeth, but the term is usually used to refer to the teeth themselves (that is, the object commonly found as fossils). A complete fossil of the "conodont animal," called Conodontephora, was discovered only in the 1980s. To add to the confusion, some researchers think the "teeth" fossils are not, in fact, teeth but a part of the head or jaw.

Corals Colonies of marine animals, or polyps, that remove calcium carbonate from seawater and deposit it around the lower parts of their bodies. New polyps grow on the old ones, thereby forming extensive limestone "skeletons." Coral reefs are simply the accumulation of these animals' skeletons plus sand and algae.

Crinoids Marine animals sometimes misleadingly called sea lilies. Their long "stems" are often preserved as fossils.

Foraminifera Single-celled, amoebalike animals with internal skeletons, most of which live on the ocean bottom. Different species live at different depths, and thus their fossils can be a clue to depth of marine sediment deposition.

Fusulinids A group of very large foraminifers that lived on shallow seafloors and were especially abundant during the Permian period. Some were several centimeters long, even though they were single-celled creatures. Because of their rapid evolution, they are helpful index fossils for Upper Carboniferous and Permian strata.

Gastropods A group of mollusks, including snails.

Stromatolites Ancient colonial algae still alive today in a few shallow seas, growing in large mounds under the water. Stromatolites precipitate layered calcium carbonate, which is preserved as one of the most widespread fossils of earth's earliest life.

Trilobites Many-legged arthropods that flourished in the early Paleozoic oceans. They were one of the first groups of animals to have hard parts and thus to be well represented in the fossil record. It is geological good fortune that trilobites both were widespread and evolved rapidly, with each species living for short periods of geologic time; they are good index fossils for the Cambrian period and some other Paleozoic rocks.

APPENDIX C

Geologic Time

The units of geologic time are crucial to understanding the discourse of earth scientists (Figure C-1).

The earth is about 4.6 billion years old. The earliest part of earth history, when the planet was still largely molten and most water existed in the gaseous phase, is termed the *Hadean eon*. We know little about this period because it left no rock record. Therefore, the oldest unit of time generally discussed in this text is the *Archean eon*, which extended up to 2.5 billion years ago. The *Proterozoic eon* followed and goes up to 543 million years. The Archean and the Proterozoic together are called the *Precambrian Eon*. During these periods, life was simple and existed only in the oceans.

The *Phanerozoic eon* began 543 million years ago and continues to the present (Figure C-2). The name comes from *phan,* meaning "visible," and *zoic,* meaning "life." During the Phanerozoic, when multicellular plants and animals evolved, life becomes easily visible in the fossil record: first fairly simple worms in the sea, then fishes, amphibians, reptiles, dinosaurs, birds, and eventually mammals, presumably followed by creatures yet to be determined. The Phanerozoic eon is important, in part because it is more recent than the Precambrian and because it tells us so much about the history of life on earth. It is broken down into the following parts:

Era	Period	Epoch
Cenozoic Era (66 million–present)	Quaternary	Holocene Pleistocene
	Tertiary	Pliocene Miocene Oligocene Eocene Paleocene
Mesozoic Era (245 million to 66 million)	Cretaceous Jurassic Triassic	
Paleozoic Era (543 million to 245 million)	Permian Pennsylvanian Mississippian Devonian Silurian Ordovician Cambrian	

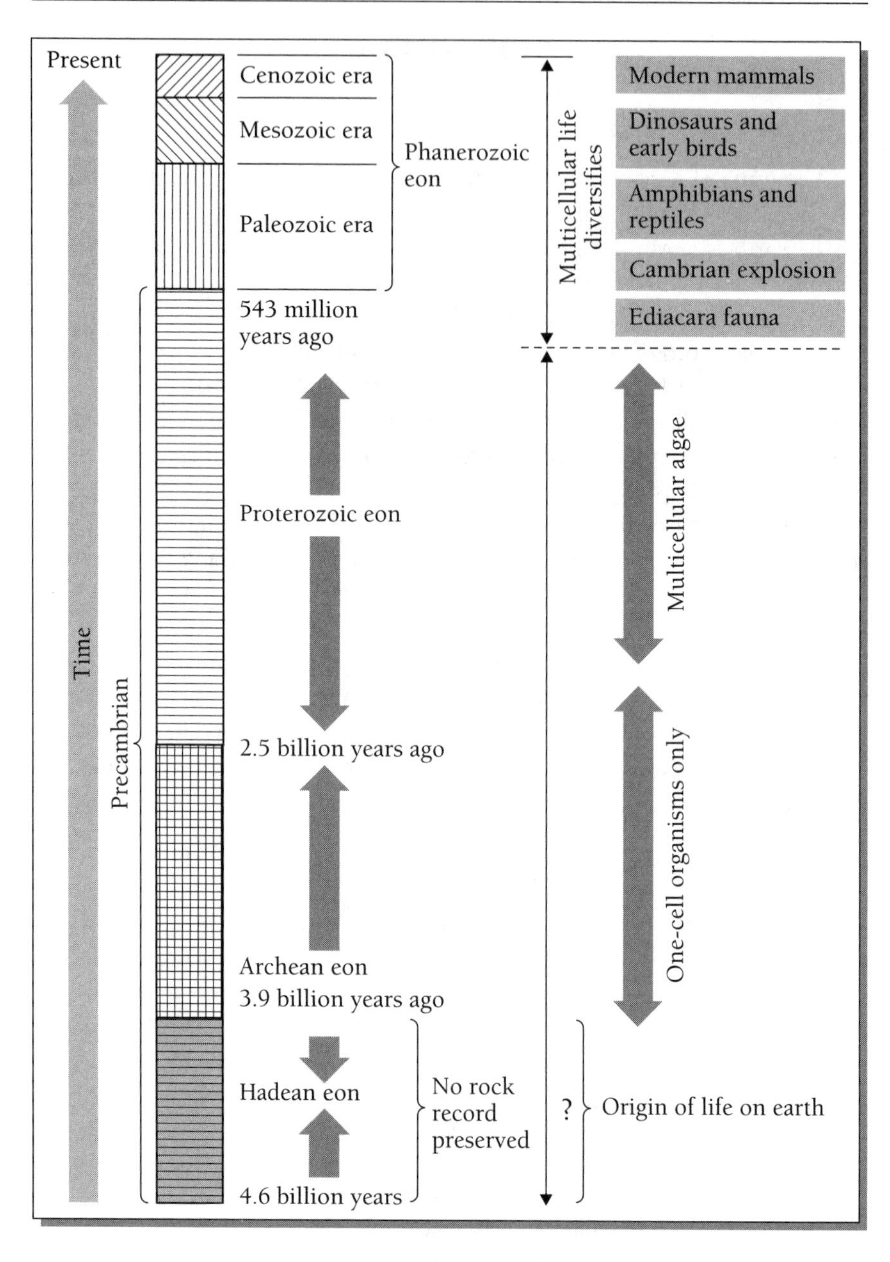

FIGURE C-1: Geologic time (note the relative length of the Precambrian).

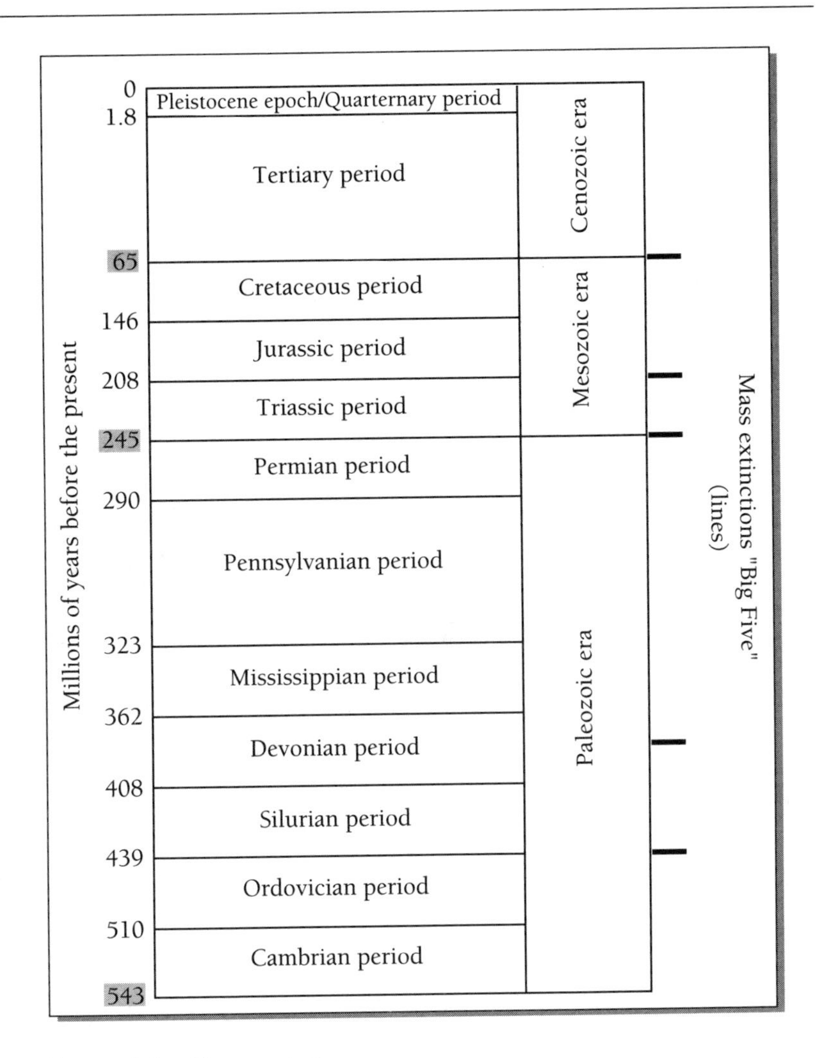

FIGURE C-2: Phanerozoic time.

THINGS TO NOTE

We know (and usually care) a lot more about the Phanerozoic eon than about the Precambrian, even though the Precambrian covers nine times as much of earth history. The Phanerozic is divided into many parts because the history of life was rapidly changing and geologists base their timescale on changes in the fossil record. The boundaries of the periods and eras are generally marked by periods of significant extinctions.

APPENDIX D

ONE APPROACH TO DEFINING A TECHNICAL TERM

Aristotle sketched a method for constructing a definition of a word, and his idea is still a fine approach to use in expressing the meaning of any scientific or technical term. He recommended first referring the term to a broader class of ideas (to show where it fits in the universe, so to speak) and then relating it to narrower concepts (to show the distinguishing features of the term).

Consider this example: Say you are asked to define the term *earth*. There's a lot you could say! It's the place where we live. It's home to dolphins and the U.S. Marine Corps. It has oceans that cover two-thirds of its surface. It moves around the sun once a year, and it turns on its axis once each 24 hours. And of course, it's the origin of the main culture of the United Federation of Planets in *Star Trek*.

BUT WHAT IS IMPORTANT FOR A DEFINITION?

Always start a definition with the term itself. For example, begin with "Earth is. . . ." Other beginnings may lead you into logical spin-offs.

Second, relate the term to a broader, more inclusive idea. In this case, the word *planet* comes to mind. Earth is one planet among many, and *planet* conveys a lot of information important to understanding Earth. So "Earth is a planet. . . ."

Third, provide information that shows how Earth can be *distinguished* from other objects in the broader class. What makes it special, a planet worthy of a particular name? "Earth is a planet that lies third from the Sun, between Venus and Mars."

Fourth, continue to give information that defines Earth's important features: "Earth is a planet that lies third from the Sun, between Venus and Mars. It is one of the terrestrial planets, composed of rocky, not gaseous, material. Earth differs from the other terrestrial planets in that it is covered by water. There is liquid water in the oceans, gaseous water in the clouds, and solid water in the polar ice caps. Earth has been the home for billions of years to many forms of life, which apparently evolved on the planet in part because of its abundant liquid water. The top portion or stony part of the Earth is divided into sections called plates that move with respect to one another. Some of these plates carry continental rock on them, and the continents protrude above the oceans as dry land. Movement of the plates gives the physical Earth a dynamic, ever-changing character . . ." *and so on as time and energy permit.*

Glossary

Advocate A person who defends a theory or an assumption is an advocate of that particular view. Scientists publish their work in the style of advocates, advancing or attacking a theory in the way that lawyers do in court. The advocacy typical of scientific discourse fits with our society's competitive and individualistic characteristics but may not be the most direct path to knowledge.

Aesthetics Aesthetics refers to such values as beauty and elegance that stand apart from pragmatic or utilitarian considerations.

Amino acids Amino acids are organic chemicals that combine to form proteins. They often are termed the "fundamental building blocks" of life on the earth, but they also are found, rather to scientists' surprise, in carbonaceous chondrites (meteorites).

Ammonia Ammonia is a compound of nitrogen and hydrogen, both of which are important constituents of life. We believe that ammonia gas was present in the earth's early atmosphere.

Anaerobic Anaerobic means without, or lacking, free oxygen. The earth's early atmosphere was anaerobic, being composed of carbon dioxide, methane, ammonia, and hydrogen sulfide gases. Some bacteria thrive in anaerobic conditions (for example, at mid-ocean ridges and in our intestines) but cannot tolerate environments containing oxygen.

Anthropic principle Some people speculate that the natural constants on which our existence seems to be contingent (and the many contingencies in the long and complex evolution of life) must result from some sort of final cause. According to this view, it is because we *Homo sapiens* are here (despite the staggering odds against our existence) that we are led to suspect final causality. This would mean that the universe has a goal or direction of which we are at least one part. Such a view can be harmonized with many religious ideas, but it is outside the realm that scientists are trained to consider.

Argument from design The theological argument that the complexity and order in nature reflects the goodness of an all-powerful creator. One branch of this idea is known as *natural theology*. James Hutton subscribed to this style of thought. The argument from design dominated much of English intellectual life until the early twentieth century.

Background extinction At least a few extinction events are always occurring in the world, an ongoing demise of species that is termed *background extinction*. Some scientists believe that episodes of so-called mass extinctions result from statistical fluctuations within the normal, background rate of extinction, just as 100-year floods are variations of normal, annual flooding. If this is so, mass extinctions do not require special or dramatic explanations.

Basalt Basalt is an igneous and volcanic rock that is relatively dense and mafic and covers the ocean floors. It forms at mid-ocean ridges and is consumed in deep marine trenches. Geologists don't know why basalt also is found in a few places on the continents, such as the Columbia Plateau basalts of Washington State.

Blue-green algae Blue-green algae are bacteria, more correctly known as *cyanobacteria*. They can be found in colonial masses such as stromatolites and are credited with supplying oxygen to the early earth's atmosphere.

Calcite Calcite is a mineral made of calcium carbonate; it is not a silicate. Seashells, limestone, and marble are made of calcite.

Carbon dioxide Carbon dioxide is a gas composed of oxygen and carbon, the familiar gas of soda pop. The early earth's atmosphere was mainly carbon dioxide. Although the current atmosphere contains only a trace of carbon dioxide, we are adding to it by burning fossil fuels. Because of this, we may be increasing the average temperatures on the earth's surface. This idea is known as the "greenhouse effect."

Carbonaceous chondrite Carbonaceous chondrite is a type of meteorite containing reduced carbon and organic components such as amino acids.

Catastrophism Catastrophism is a theory about earth history that emphasizes rapid change on a large scale. The impact of meteorites and the outburst flooding in Washington State during the last ice ages are in accord with catastrophism. Nevertheless, established geology was fully gradualistic by the end of the last century, and respected geologists resisted catastrophism well into the middle of this century. But now, in the modern concept of uniformitarianism, geologists make room for catastrophic events.

Chaos theory In recent years, scientists have begun to study systems in which tiny differences in state (perturbations) create, over time, substantial differences in the system. For example, the weather is a natural system that we have (finally) admitted must be studied in this manner.

Coacervates Coacervates are small spheres that can be formed from proteins and that possess many, but not all, of the properties of life. They may have been important to the origins of life on the early earth.

Continental crust Continental crust is the rock making up the continental landmasses and includes rocks that are several billion years old. It is thicker and less dense than the lower-lying oceanic crust and is composed mainly of granite, which is a felsic, plutonic rock.

Continental drift Continental drift is a wide-ranging geologic theory proposed by Alfred Wegener. Although he did not have a clear picture of the mechanism that moved the continents and did not have information about the ocean basins, he was sure the continents had split apart and moved across the globe. For many decades, continental drift was dismissed by established geologists, but it eventually was incorporated into the modern theory of plate tectonics.

Contingency If an event's occurrence (or nonoccurrence) depends on a prior event, it is contingent on it. For example, a student's graduation from college is contingent on fulfilling the science requirement.

Convection (in the earth) Convection within the earth was assumed in the late 1960s to drive the tectonic plates. It was the process by which hot, molten parts of the mantle rise toward the surface at mid-ocean ridges and sink at convergent margins. Geologists no longer believe that the inner earth truly convects.

Cosmic dust Dust particles in space are known as *cosmic dust*. Although they fall to the earth at a slow rate, these dust particles can be important to the geochemical budgets of some metals, including iridium. Cosmic dust is relatively rich in iridium.

Coulee Coulee is a term used by the residents of eastern Washington State to denote the wide, flat-bottomed channels of the Scablands. Viewed from the air, the coulees form a braided pattern. J Harlan Bretz recognized them as resulting from catastrophically deep floodwaters.

Craton Craton refers to the relatively stable and rigid part of the continental crust that was formed in the earth's early history. Cratons apparently grow by acquiring younger, accreted terrains on their margins.

"Creation scientists" "Creation scientists" are a small minority of Christians who subscribe to a literal interpretation of the Genesis account of the earth's creation. This group should be known as *biblical literalists*. These people hold that approximately 6000 years ago, God created the world in a series of miracles. According to this view, all sedimentary rocks and fossils were created during Noah's flood. "Creation science" is not scientific but, rather, relies on scripture as its sole source of authority.

Darwinism Charles Darwin advanced a theory of biological evolution dependent on natural selection, sometimes called *descent through modification*. His theory marks the first time that biologists had an explanation for the regularities observed in taxonomy. Neo-Darwinism and punctuated equilibrium are modern descendants of classical Darwinism.

Dendritic This term refers to any natural pattern of branching which leads to further branching, like a tree. Normal drainage patterns are called dendritic because small creeks feed into larger and larger ones.

Descriptive science A physical science (like chemistry or physics) strives to be predictive, whereas a descriptive science (like geology) is devoted mainly to describing rocks, minerals, faults, and other features of the earth. Unlike physical science, geology is usually not predictive, but it can explain what is found in the rock record.

DNA Along with RNA, DNA, a complex organic molecule, enables cells to store and use genetic information. It is present in cells in all life-forms on earth.

Dualism Dualism is the assumption that all existing entities are either material or immaterial. Most of Western civilization has assumed dualism (for example, a material brain but an immaterial mind, a material body but an immaterial spirit). Whereas science assumes simple materialism, most religious members of our society are, in some sense, dualists.

Efficient causality An immediately preceding and naturalistic cause for an event is the hallmark of efficient causality. We are used to assuming this kind of cause for events in the physical (nonliving) world around us.

Empirical Empirical refers to something that is dependent on experiment or observation. The facts to which empirical work appeals are in the "public domain" in the sense that anyone can verify them. Empirical research is characterized by a particular style of logic and reasoning, although it (at least as practiced by scientists) is not free of individual preference and personal idiosyncracies. Some philosophers maintain that no clear line can be drawn between the empirical realm and personal and intellectual values.

Endothermic Endothermic means "warm-blooded." Contrary to earlier belief, many scientists now argue that some dinosaurs were endothermic and therefore capable of long-term, fast movements, just as mammals are.

Establishment of religion clause The courts have interpreted this clause in the First Amendment to the U.S. Constitution as separating various facets of religious belief from government programs (such as public school curricula). Because our society is composed of people belonging to different religions, our Founding Fathers prohibited the state from "establishing" a particular religion.

Eukaryote Eukaryotic cells have a nucleus, and their DNA is organized in chromosomes. Such cells are more advanced than prokaryotes.

Evolution Evolution can refer to any change in a system, but it generally applies to one or more scientific theories concerning the history of life.

Extraterrestrial Any body or substance external to, or outside, the earth is extraterrestrial (remember ET?).

Falsification Falsification is the process of gathering evidence that could disprove a theory. Some philosophers argue that all scientific claims must be, at least in principle, falsifiable, but this has not been the case throughout the history of science.

Feldspar The feldspar minerals are the most common in the earth's crust. They are made of silica and a few other elements. Feldspars weather more rapidly than quartz.

Felsic Felsic rocks are less dense than mafic rocks and contain greater proportions of silica and alumina. Granite is a felsic rock making up much of the continental crust, which is therefore less dense than the basaltic oceanic crust.

Final causes A final cause is dependent on the future or some goal or outcome. It is an idea congenial to various religious doctrines, in contrast to efficient causes. It also seems to accord with our personal experiences of choice and will.

Garnet Garnet is a silicate mineral formed in a metamorphic environment and is found in schists and gneisses. It is a semiprecious gem resistant to weathering.

Gene A gene is part of the chromosome and encodes the characteristics passed from one generation to the next.

Geosyncline According to the geologic theory proposed by J. D. Dana and his colleagues, geosynclines were regions of the crust that had undergone extensive vertical displacement. Mountain chains were geosynclines that had been downwarped and then raised up. In the theory of plate tectonics, the strong sense of vertical displacement was exchanged for lateral movement.

Glacial erratics Glacial erratics (or erratic boulders) are boulders that glaciers have carried far from their original site.

Gneiss Gneiss is a high-grade metamorphic rock derived from either shale or granite and is identified by its swirly, soupy texture.

Gradualism Gradualism is a framwork of thought opposed to catastrophism. Gradualism holds that the important processes of earth history are gradual, and during the last century, it was thought to be the natural complement of uniformitarianism.

Granite Granite is a felsic plutonic rock composed of quartz, two kinds of feldspar, and minor mafic minerals. It is a familiar rock, often used as building stones. As a rock made of interlocking crystals that fill all spaces, granite's texture betrays its igneous origin.

Graphite Graphite is composed of pure carbon and is found in the oldest sedimentary rocks of the Archean eon.

Gravel bars Gravel bars are elongate mounds of gravel laid down by rivers or floodwaters. In the Channeled Scablands, one "mega" gravel bar extends for 10 miles.

Hard parts Hard parts are bones, shells, and other animal parts that are durable enough to withstand some abrasion. They are much more apt to be preserved as fossils than are soft parts such as flesh, hair, or feathers.

Heat flow The core of the earth is hot, and heat is always flowing outward toward the surface of the planet. More heat is added to this flow by radioactive decay, especially in the crust. In some places, heat flow is unusually high, such as at active volcanoes and mid-ocean ridges.

Historical science A historical science like geology cannot test many of its propositions with conclusive experiments. Rather, geologists try to establish a narrative account of events that fits with the empirical evidence.

Hydrogen sulfide Hydrogen and sulfur combine to form hydrogen sulfide gas, which geologists believe was present in the early earth's atmosphere.

Hydrothermal Hydrothermal is a term from the Greek words for "water" and "heat." Hot water circulates in some parts of the crust, and often it is saline and carries dissolved constituents, including metals that can form ore deposits. Hydrothermal systems also form hot springs and geysers, such as those found in Yellowstone National Park.

Hypothesis A hypothesis is an idea or conjecture about the explanation of a set of data or observations. Scientists use the word to mean something weaker than a theory, and much weaker than a law.

Ice-drop pebbles/cobbles/boulders Rocks may be carried within and atop glacial ice that is swept along by floodwaters. When the ice melts sufficiently, the rock particle is "dropped," usually into much finer sediment. In the Scablands, geologists often find isolated cobbles of Montana rocks in fine sediment in the bottom of coulees.

Igneous rocks All rocks formed from molten material are igneous. Basalt, obsidian, and granite are igneous rocks. Mafic igneous rocks are denser than felsic igneous rocks and contain more iron and magnesium.

Information theory Information theory is concerned with the structure of information. Rich information systems, such as the complex coding of DNA, are the opposite of random noise and are necessary for life on our planet.

Iridium anomaly Iridium is a trace element in the earth that was discovered in relative abundance in the thin clay boundary between the latest Cretaceous and the earliest Tertiary (that is, the K-T boundary). Sources of this iridium might be cosmic dust falling to the earth, mafic volcanism on the planet, or the impact of a meteorite.

Isostasy The theory of isostasy, developed in the last century, states that the earth's crust "floats" on a nearly liquid layer underneath. Because the continental crust is thick but less dense, it therefore floats higher than the thinner and more dense oceanic crust. Isostasy was incorporated into plate tectonic theory.

K-T boundary The transition from the Cretaceous to the Tertiary (that is, from the Mesozoic to Cenozoic eras) is represented for convenience by the phrase K-T boundary. The nature and cause of the extinctions at the K-T boundary are still being debated.

Lake Bonneville During the last ice age (in the Pleistocene), a large lake covered much of Utah. At some point, it downcut its banks in southern Idaho, causing a catastrophic flood down the Snake River drainage.

Land bridges Land bridges were thought to have existed in the geologic past between, for example, South America and Africa. They help explain similarities among animals and fossils on different continents. Later, in the more recent past, these bridges were believed to have sunk into the oceanic crust.

Latent plastic virtue Latent plastic virtue was the creative life force in the inorganic world that Aristotle and others hypothesized was responsible for the (ongoing) spontaneous generation of life and possibly for the patterns in figured stones.

Limestone Limestone is a sedimentary rock made of the mineral calcite. Although limestone can form through inorganic processes, it usually is created by biological activity. Sea creatures secrete calcite to form their shells. If these shells accumulate on the ocean floor and become stone, they form fossil-rich limestone.

Loess Loess is wind-blown dust that accumulates in thick deposits. The basalt in the Channeled Scablands is crowned by rich loess hills in those places that escaped flooding during the Pleistocene.

Macroevolution Macroevolution is the Darwinian idea that not only could individuals within a species vary, but also species themselves had the capacity to change and to generate multiple new species over time.

Mafic Mafic igneous rocks are dark colored and contain more iron and magnesium than felsic igneous rocks do.

Marble Marble, which is composed of calcite, is a metamorphic rock derived from limestone. Sculptors often use marble for their work.

Margins (in plate tectonics) Margins are the boundaries between the plates of the outer earth. There are three kinds of margins *Convergent margins* are areas in which two tectonic plates are moving toward each other. If two pieces of continental crust collide, mountains like the Himalayas will be the result. Trenches in the ocean also mark convergent margins. *Divergent margins* are areas in the earth's crust where two tectonic plates are moving away from each other. The mid-ocean ridges are divergent margins. *Transform margins* are areas in the earth's crust in which two tectonic plates are moving past each other. The San Andreas Fault displays this kind of transform movement.

Marine sedimentary record Marine environments usually create continuous sedimentary deposits, which are the most complete record we have of the history of life. Naturally, these deposits preserve only marine (ocean) plants and animals. Terrestrial (continental) organisms are much less frequently preserved than are marine life-forms.

Mass extinction The history of life appears to some researchers to have undergone periods of abrupt and mass, or wide-ranging, extinctions. The five greatest such events are known as the *Big Five*.

Materialism Materialism (or naturalism) is the assumption that all existing entities are material. Many modern intellectuals are materialists, and science limits itself to phenomena in the material world.

Metamorphic rocks Metamorphic rocks are formed at increased temperature and pressure and often with the aid of chemically active fluids. Sedimentary and igneous rocks may be metamorphosed within the earth to create such rocks as schist, gneiss, marble, and slate.

Meteorite Meteorites are extraterrestrial bodies that fall to the earth. They are composed of silicate (rocky) material, metals (principally iron and nickel), or reduced carbon (for example, carbonaceous chondrites containing amino acids and other organic compounds).

Methane Carbon and hydrogen form methane, the principal component of natural gas. Methane is believed to have been present in the early earth's atmosphere.

Method of multiple working hypotheses The method of multiple working hypotheses was an approach to scientific research advocated by T. C. Chamberlin, who assumed that multiple hypotheses would spring directly from the physical world into the mind of a scientist. The researcher, said Chamberlin, should try to be impartial and develop evidence for each hypothesis.

Methodology The research methods in scientific disciplines constitute their professional methodology. The principle of parsimony and the assumption of efficient causality are part of the methodological assumptions of all scientific research.

Mica Mica is a silicate mineral characteristic of igneous and metamorphic rocks. It is distinguished by the flat sheets into which it breaks. Both white (muscovite) and dark (biotite) mica occur in nature.

Microevolution Changes that occur within a species are examples of microevolution. A new species is not created in this process. Both "creation scientists" and scientists can agree on examples of microevolution, including specialized breeds of dogs and horses, and bacteria that have become highly resistant to antibiotic agents.

Mid-ocean ridges (**MOR**) Mid-ocean ridges are linear ridges in the middle of the oceans. They were mapped after World War II and were discovered to be zones of high heat flow and high seismicity. Magma (molten rock) wells up from the mantle at mid-ocean ridges and forms basaltic, oceanic crust which spreads slowly away from the ridges in both directions.

Minerals Minerals are naturally occurring crystalline substances with a definite chemical composition. Aggregates of minerals form rocks.

Mutation Random genetic change, caused by mutation, leads to variations in a population. According to Darwinism, natural selection slowly acts on all variational differences to create new species.

Narrative science Geology is a historical science embracing the art of narration in all its major theories. Unlike physics, the evidence of interest to geologists is embedded in time. Geologists sometimes refer to their hypotheses as *stories*.

Naturalism Naturalism (or materialism) is the assumption that everything is made of material objects (atoms and electrons, not spirits and disembodied minds).

Neo-Darwinism This variation of Darwinism emphasizes that natural selection acts on variability inherited through genes and encoded in the DNA molecule. Neo-Darwinism is a gradualistic theory.

Neptunism Neptunism was the idea held by some early geologists that almost all rocks formed at the bottom of an ancient sea or during the worldwide flood of Noah's time.

Niche A biological niche is a position in the ecosystem occupied by an organism or organisms.

Noise Noise is an informal term used by scientists to denote data or observations that do not follow any pattern but, rather, appear random.

Ockham's razor Ockham's razor is an informal reference, frequently used by scientists, to the principle of parsimony. William of Ockham was a medieval philosopher who argued that God created the earth in a manner that reflects the divine perfection demonstrated by simplicity. Accordingly, Ockham believed that a simple argument was preferable to a complex one.

Ozone Ozone is a form of oxygen gas composed of three, rather than two, oxygen atoms. It is highly reactive and is present in smog and also in the upper atmosphere where it reacts with ultraviolet light, effectively screening the globe from the harmful effects of such radiation. Unfortunately, the ozone in smog does not move up into the highest layers of the atmosphere but reacts with compounds nearer the surface.

Pangaea Pangaea is the name of the supercontinent that Alfred Wegener posited in his theory of continental drift. The term also is used in plate tectonics.

Paradigm Paradigm is a term used by Thomas Kuhn that has become commonplace in many different disciplines. It refers to a general model or framework of thought. The German word *Weltanschauung* (worldview), used in philosophy, is roughly equivalent.

Philosophical materialism Philosophical materialism holds that the universe contains only material substances and natural causation. Everything, therefore, can be addressed through scientific methodology. Scientists are often personally committed to this philosophy, although individual scientists have always felt free to disagree on the basis of personal religious conviction or intellectual skepticism. Philosophical materialism is not in accord with religious beliefs or mystical experiences, and to some observers, its explanation for aesthetic and moral values, as well as human choice, appears strained.

Plate tectonics Plate tectonics encompasses current geophysical and geological theories concerning the earth. It combines ideas found in continental drift and seafloor spreading to account for the patterns of seismicity, petrology, and topography that geologists see on the continents and in the ocean basins.

Plutonic Igneous rocks that form from a magma deep within the earth are termed *plutonic* and are characterized by large, interlocking crystals.

Principle of parsimony The principle of parsimony is one of the methodological assumptions of scientific research, recommend the simplest possible explanation of natural phenomena. William of Ockham, a medieval theologian and philosopher, is often credited with first stating the principle, although he had in mind something quite different from what secular scientists imagine.

Prokaryote A cell that does not have a nucleus and in which the DNA is not organized into chromosomes is a prokaryote. Such cells are considered primitive. Bacteria of the early earth were prokaryotes, but later, more advanced life has been made up of eukaryotic cells.

Pseudoextinction If an organism changes very gradually through many generations and eventually evolves into a new species, the original organism undergoes only a pseudoextinction, since much of its genetic material is represented in the new species. Darwin assumed that most extinctions in the history of life were pseudoextinctions. He was committed to this idea because it was in accord with gradualism.

Punctuated equilibrium Punctuated equilibrium is one strand of current theory regarding the evolutionary history of life on earth. It fits relatively well with the fossil record, emphasizing long periods of stasis punctuated by speciation events that allow rapid evolutionary change.

Quartz Quartz is a mineral made of pure silica (one silicon atom bound to four oxygen atoms). Quartz is the principal element in white sand because it is resistant to weathering.

Quartzite When sandstone is metamorphosed, it forms quartzite. The quartz grains are recrystallized to form one interlocking mass of silica crystals. Quartzite is highly resistant to weathering.

Radiometric dating The radioactive elements present in some minerals and rocks allow geologists to date earth materials quantitatively (that is, absolutely). However, the geologic timescale was based on the principles of *relative* time alone, without the benefit of radiometric dating, although absolute dates have now been appended to each period of the timescale.

Rational For a thought to be rational, it must be in accord with known evidence and with reason. Some thoughts can be nonrational without being irrational. For example, an art critic's aesthetic value judgments of certain Impressionist paintings may be nonrational but not irrational.

Reduced Reduced refers to atoms that have lost electrons from their orbitals. Oxygen can be an agent of reduction. Reduced carbon, nitrogen, and hydrogen all are required for life as we know it. The earth's early history reflects a reducing atmosphere.

Rhythmites Rhythmites are graded sedimentary beds, that is, beds with coarse material on the bottom and finer particles toward the top. Such beds, geologists believe, are formed from floods in which the first particles to settle out are coarse and not until later do the finer particles sink to the bottom. Rhythmites were one piece of evidence that J Harlan Bretz used to argue for the flooding of the Channeled Scablands.

Ripple marks Ripple marks are the roughly parallel ridges of sediment preserved in sedimentary rocks and deposits. They allow geologists to determine which way was "up" at the time of sedimentary deposition. When enormous ripple marks were documented from the air near Missoula, Montana, they were recognized as the source of the catastrophic flooding proposed by J Harlan Bretz.

RNA RNA, like DNA, is a complex organic molecule that cells use to store and transmit genetic information by synthesizing the chromosomes' protein.

Rock cycle Geologists organize the basic types of rock into a cycle that has existed throughout earth history. Molten material cools and forms igneous rocks. Weathering and erosion lead to the formation of sedimentary beds that can be transformed into stone. Elevated temperatures and pressures can convert either igneous or sedimentary rocks into metamorphic rocks.

Sandstone Sandstone is a sedimentary rock formed from sand. The sand usually is quartz and often is of marine origin. Occasionally, however, desert sand dunes (above sea level) can be covered, buried, and lithified to create nonmarine sandstone.

Scale Although the concept of scale or size is simple, it determines what observations, and therefore what conclusions, scientists interested in a given problem may make. Only a whole-earth perspective made possible the theory of plate tectonics. Similarly, J Harlan Bretz's main contribution to field science was stressing the large-scale sedimentary features that catastrophic flooding could cause.

Schist Schist is a metamorphic rock derived from shale, and it contains mica minerals, which give the rock a distinctive sheen or luster.

Seafloor spreading Seafloor spreading was a theory advanced in the early 1960s to explain the topography of the mid-ocean ridges, arguing that they were spreading centers for basaltic magma welling up and forming new crust. This theory fed into the thinking that created plate tectonic theory just a few years later.

Sedimentary rocks Sedimentary rocks form through the aggregation of minerals and preexisting rocks or through chemical or biological precipitation. Sandstone, limestone, shale, and conglomerate all are sedimentary rocks. Sedimentary rocks may contain fossils, whereas igneous and metamorphic rocks do not.

Seismic Seismic refers to earthquakes. Applied physicists (geophysicists) study seismic waves for information about the structure of the rocks beneath the earth's surface. Oil companies explore for new oil fields using sophisticated seismic methods.

Selective advantage If an organism has a feature different from its fellows that helps it survive and reproduce, that feature is a selective advantage. Darwinism and more recent theories state that selective advantages are what drives evolutionary change.

Separate creation From the ancient world through the last century, people assumed that *Homo sapiens* were the result of some special act of creation, different in kind from the events that led to animal life.

Shale Shale is a common sedimentary rock made of muds and clays. It is dark and breaks most easily in one direction.

Slate Slate is a flat and dense product of the metamorphism of shale. It is used in (old) blackboards and as the base for pool tables.

Speciation Speciation is the creation of a new species, although it remains poorly understood despite more than a century of Darwinian and neo-Darwinian investigations. The evolutionary theory called *punctuated equilibrium* asserts that most evolutionary change occurs during speciation events.

Spontaneous generation From ancient times until the last century, most people believed that simple forms of life arose from inorganic components when the conditions were propitious. The creative "life force" responsible for spontaneous generation was attacked by Louis Pasteur, who experimentally demonstrated that in sterile conditions, life did not arise from inorganic components.

Stasis The term *stasis* springs from the evolutionary theory called *punctuated equilibrium*. Stasis periods are the long passages of time during which an organism does not significantly change, whereas speciation events are periods marked by rapid evolutionary change.

Stromatolites Stromatolites are mounds of colonial, blue-green algae (cyanobacteria) that thrived in the shallow marine environments of the Archean and Proterozoic eons. A few stromatolites can still be found where other marine life is limited. Stromatolites are believed to have been one of the principal agents of the oxygenation of the Precambrian earth's atmosphere.

Taxonomists Taxonomists are biologists who specialize in classifying plants and animals into related groups and families. Taxonomy was a successful descriptive science; Charles Darwin tried to explain the regularities of taxonomy using his theory of evolution.

Topography The earth's surface features—its highs and lows, valleys and mountains, plains and abysses—constitute its topography. The topography of mid-ocean ridges and of oceanic trenches are explained by plate tectonic theory.

Ultraviolet (UV) Ultraviolet radiation—often termed ultraviolet light, although it is invisible—is radiation of a higher frequency than what our eyes can detect. Ultraviolet radiation is part of the energy that our planet receives from the sun. Although it is harmful to life, much of the UV radiation is absorbed by the ozone layer in the earth's upper atmosphere, which is why the newly discovered "ozone hole" in the atmosphere above Antarctica has so alarmed some scientists.

Uniformitarianism Uniformitarianism is a difficult term; it is still used inconsistently by professional geologists. The word reflects the idea that geologists should study currently active processes to gather information about the rock record. Geologists sometimes use the phrase "The present is the key to the past" to sum up this idea. Notice, however, that geologists have no reason, other than their own convenience, to assume that the present is representative of earth history. Rather, they make this assumption as one instance of the principle of parsimony, as it seems simpler to think that nature worked in the past as it does now than to think otherwise.

In the last century and the earliest part of this century, uniformitarianism was interchangable with gradualism. Gradualism was an important substantive claim about the earth with which all respected geologists agreed. This is, however, no longer the case, for geologists now recognize that catastrophic processes have been important to earth history. This understanding is sometimes called the *new uniformitarianism*.

Varves Varves are small, paired, light and dark layers in a sedimentary deposit that show variation in sedimentation within one year.

Volcanic Volcanic rocks are igneous rocks that form at or near the earth's surface. They generally contain small mineral crystals, apparently because rapid cooling is not conducive to the growth of large crystals.

Volcanism Volcanism—and later, plutonism—stood opposed to Neptunism. Volcanists held that basalt was of volcanic, not sedimentary, origin. Volcanists posited a molten mass of high-temperature liquid rock beneath the earth's crust.

Annotated Bibliography

A General Note

Many popular-level articles on the topics discussed in this book are available in such magazines as *Natural History, Earth,* and *Scientific American.* Collections of essays by Stephen Jay Gould are available in virtually all libraries, and Richard Dawkin can be consulted for a view of evolutionary theory different from Gould's. Dawkin's book *The Blind Watchmaker* is a classic advocacy piece. The theistic arguments against recent evolutionary theories are occasionally published by Intervarsity Press (Illinois), but the quality of the work varies. Publications from the Creation Science Institute (California) are antiempirical and, indeed, anti-intellectual.

The award-winning PBS series *Nova* also treats many different geologic issues, and videotapes of that series are available through many public libraries and universities. The style of geologic debate may be most easily seen in the interviews with researchers who are currently struggling to better understand earth processes and earth history.

Broader scientific and philosophical issues are well reviewed in a series of best-selling volumes by Paul Davies. Frank Tippler's *Physics and Immortality* should also be useful to interested readers, but it is much more advanced than Davies's popular-level works. Timothy Ferris has written several mind-expanding books on issues of science, the mind, and physics, including *Coming of Age in the Milky Way*.

The social, political, and legal consequences of some of the conflicts described in this book were recently summarized from a religious point of view in Stephen Carter's *The Culture of Disbelief* and Phillip Johnson's *Darwin on Trial.*

Chapter 1

The research project described in Chapter 1 resulted in a number of publications, including the following:

Fehn, U., E. K. Peters, S. Tullai-Fitzpatrick, P. W. Kubik, P. Sharma, R. T. D. Teng, H. E. Grove, and D. Elmore. "^{36}Cl and ^{129}I Concentrations in Waters of the Eastern Clear Lake Area, California: Residence Times and Source Ages of Hydrothermal Fluids." *Geochimica et Cosmochimica Acta* 56 (1991): 2069–2079.

Peters, E. K. "D-^{18}O Enriched Thermal Waters, Coast Range, Northern California." *Geochimica et Cosmochimica Acta* 57 (1993): 1093–1104.

———. "Gold-Bearing Hot Spring Systems of the Northern Coast Ranges, California." *Economic Geology* 86 (1991): 1519–1528.

———. "Wilbur Hot Springs and the McLaughlin Mine." Paper presented to the annual meeting of the Geothermal Resources Council and published in *Transactions* 15 (1991): 41–46.

Chapter 2

Barber, Bernard. "Resistance by Scientists to Scientific Discovery." *Science* 134 (1961): 596–602. In this article, Barber (a social scientist) explains scientists' resistance to new ideas and their reasons for maintaining such attitudes.

Chamberlin, T. C. "The Method of Multiple Working Hypotheses." *Science* 15 (1890): 92–96, reprinted in *Science* 148 (1965): 754–759. This article, although philosophically naive, is still assigned in many geology classes and is still uncritically accepted by graduate students and professors alike.

Davies, Paul. *The Mind of God: The Scientific Basis for a Rational World*. New York: Simon & Schuster, 1992. Davies is a physicist and an excellent writer. I recommend all his books, and each has something to say about the themes of this book.

Davis, Bernard D. *Storm over Biology: Essays on Science, Sentiment, and Public Policy.* Buffalo, NY: Prometheus Books, 1986. This collection of essays favors the objectivity of empirical knowledge. Davis attacks Lewontin and other scientists in biology and biochemistry who argue for the subjective nature of science. Davis distinguishes between the scientific method (which he admits has many subjective components) and scientific knowledge (which he sees as objective and corresponding to the facts of the natural world). He defends the basic notions of sociobiology, and unlike Gould, he believes that science can usefully comment on values and morality.

Edey, Maitland, and Donald Johanson. *Blueprints: Solving the Mystery of Evolution*. Boston: Little, Brown, 1989. These two scientists discuss current evolutionary theory.

Ferguson, Kitty. *The Fire in the Equations: Science, Religion and the Search for God*. Grand Rapids, MI: Eerdmans, 1995.

Feyerabend, Paul. *Against Method*. London: New Left Books, 1975.

———. *Science in a Free Society*. London: New Left Books, 1978. Feyerabend was a philosopher and devoted much of his energy to criticizing the cult of empiricism which, he felt, grips this society. His books should be read by anyone with an interest in methodological and philosophical issues of science.

Gustafsson, Bengt. *The New Faith–Science Debate*. Minneapolis: Fortress Press, 1989. Gustafsson is a Swedish physicist with interests in religion and philosophy. Fortress Press, it should be noted, is a religious publishing house.

Holton, G. *Science and Anti-science*. Cambridge, MA: Harvard University Press, 1993. An advanced survey of the issues.

Hull, David L. *Science as a Process*. Chicago: University of Chicago Press, 1988. This interesting volume touches on many subjects and tries to be realistic about what constitutes scientific work.

Kuhn, Thomas. *The Structure of Scientific Revolutions*. Chicago: University of Chicago Press, 1962. This book marks the beginning of a major debate in the history of science about the mode and tempo of scientific progress.

Lewontin, R. C. *The Doctrine of DNA: Biology as Ideology*. Harmondsworth: Penguin Books, 1991. Lewontin is on the end of several continua concerning scientific apologetics. Biologists and anyone interested in the subjective nature of science should read this book.

Moore, John A. *Science as a Way of Knowing*. Cambridge, MA: Harvard University Press, 1993. This volume is a solid survey of how scientific knowledge moves forward.

Thurman, L. Duane. *How to Think About Evolution*. Madison, WI: Intervarsity Press, 1978. This is a creationist book remarkably generous to the scientific method; the author does not understand the human and subjective nature of scientific research.

Chapter 3

Adams, Frank D. *The Birth and Development of the Geological Sciences*. New York: Dover, 1938. This is an old book, but it has good coverage of the nineteenth century.

Allen, John E., and Margorie Burns, with Sam C. Sargent. *Cataclysms on the Columbia*. Portland, OR: Timber Press, 1986. This is a popular-level book, quite readable and fun. It is a good introduction to the Channeled Scablands of Washington State.

Drake, E. T., and W. M. Jordan, eds. *Geologists and Ideas: A History of North American Geology*. Centennial special vol. 1. Washington, DC: Geological Society of America, 1985.

Frodeman, Robert. "Geological Reasoning: Geology as an Interpretive and Historical Science." *Geological Society of America Bulletin* 107 (1995): 960–968. Professor Frodeman argues that geology and geological research methods are the queen of the empirical realm.

Greene, Mott T. *Geology in the 19th Century*. Ithaca, NY: Cornell University Press, 1982. A good summary.

Gustafsson, Bengt. *The New Faith–Science Debate*. Minneapolis: Fortress Press, 1989. See the annotation for this book in the references for Chapter 2. An easy-to-read summary.

Harrington, John W. *Dance of the Continents: Adventures with Rocks and Time.* Boston: Houghton Mifflin, 1983. This is an introduction to elementary geology.

Hull, David L. *Science as a Process*. Chicago: University of Chicago Press, 1988. See the annotation for this book in the references for Chapter 2.

Leet, Don L., and Sheldon Judson. *Physical Geology*. 1st ed. New York: Prentice-Hall, 1954. For several decades, numerous editions of Leet and Judson's geology textbooks were among the best selling in the United States.

Moore, John A. *Science as a Way of Knowing.* Cambridge, MA: Harvard University Press, 1993.

Officer, C., and J. Page. *Tales of the Earth.* New York: Oxford University Press, 1993. This is a fine contribution, written at the popular level and recommended to anyone who has enjoyed the geological sections of my book.

Schumm, S. A. *To Interpret the Earth: Ten Ways to Be Wrong*. Cambridge: Cambridge University Press, 1991.

Von Bubnoff, S. *Fundamentals of Geology*. Edinburgh: Oliver and Boyd, 1963. Von Bubnoff shows more philosophical sophistication than many geologists, perhaps owing to his German education.

Chapter 4

Gohau, G. *A History of Geology.* New Brunswick, NJ: Rutgers University Press, 1991. This book clarifies the French contribution to geology.

Hallam, A. *Great Geological Controversies*. Oxford: Oxford University Press, 1983. This is an excellent history of many of the debates concerning Neptunism, volcanism, and the origins of basalts and granites.

Press, F., and R. Siever. *Understanding Earth*. New York: Freeman, 1994. This best-selling geology textbook explains minerals and rocks in much greater detail than this book does.

Chapter 5

The accomplishments of the three great geologic gradualists are recorded in the following works:

Hutton, James. *The System of the Earth, Its Duration, and Stability.* Edinburgh: Royal Society of Edinburgh, 1785.

Lyell, Charles. *Prejudices Which Have Retarded the Progress of Geology*. London: Murray, 1872.

Playfair, John. *Illustrations of the Huttonian Theory of the Earth*. Edinburgh: Cadell & Davis, 1802.

Excerpts from each of these works may be found in Claude C. Albritton Jr., ed., *Philosophy of Geohistory, Benchmark Papers in Geology*. Vol. 13. New York: Halstead Press/Wiley, 1975.

Other Sources

Dalrymple, G. B. *The Age of the Earth*. Stanford, CA: Stanford University Press, 1991. This excellent book gathers together a variety of evidence regarding the earth's enormous age.

Gould, Stephen Jay. "Toward the Vindication of Punctuated Change." In *Catastrophes and Earth History*. Ed. W. A. Berggren and John Van Covering. Princeton, NJ: Princeton University Press, 1984.

Leet, Don L., and Sheldon Judson. *Physical Geology*. 1st, 2nd, and 3rd eds. New York and Englewood Cliffs, NJ: Prentice-Hall, 1954, 1958, and 1965.

Lindley, David. *The End of Physics: The Myth of a Unified Theory*. New York: Basic Books/HarperCollins, 1993.

Shea, James. "Twelve Fallacies of Uniformitarianism." *Geology* 10 (1982): 455–460. An excellent and accessible article.

Wigner, Eugene P. "The Unreasonable Effectiveness of Mathematics in the Natural Sciences." *Communications on Pure and Applied Mathematics* 13 (1960): 1–14. This is a surprisingly readable article, in which the author argues that the useful character of mathematics is something "bordering on the mysterious."

Chapter 6

Alden, W. C. "Discussion: Channeled Scabland and the Spokane Flood." *Washington Academy of Science Journal* 17 (1927): 203. This work attacks Bretz's hypothesis of catastrophic flooding.

Allen, John, and Marjorie Burns, with Sam Sargent. *Cataclysms on the Columbia*. Portland, OR: Timber Press, 1986. A popular and excellent volume.

Baker, Vic. "Surprise Endings to Catastrophism & Controversy on the Columbia." *GSA Today* 15 (1995): 170–172.

Bretz, J Harlan. "The Channeled Scablands of the Columbia Plateau." *Journal of Geology* 31 (1923): 617–649.

———. "Glacial Drainage on the Columbia Plateau." *Geological Society of America Bulletin* 34 (1923): 573–608.

———. "The Lake Missoula Floods and the Channeled Scabland." *Journal of Geology* 77 (1969): 505–543. In these (and many other) articles, Bretz sets out his catastrophic theory of outburst flooding.

Gilbert, G. K. "The Ancient Outlet of Great Salt Lake." *American Journal of Science* 15 (1878): 256–259.

———. "Lake Bonneville." *U.S. Geological Society Monograph* 1 (1890).

———. "The Scientific Ideas of G. K. Gilbert." Ed. Ellis Yochelson. *U.S. Geological Society Social Paper* 183 (1980).

It seems clear that if Gilbert and Bretz had been able to do their fieldwork together, we might have been able to understand Pleistocene catastrophic flooding much sooner.

Gilluly, J. "Discussion: Channeled Scabland and the Spokane Flood." *Washington Academy of Science Journal* 17 (1927): 203–205. This work attacks Bretz's hypothesis of catastrophic flooding.

Gould, Stephen Jay. "Toward the Vindication of Punctuation Change." *In Catastrophes and Earth History.* Ed. W. A. Berggren and John Van Covering. Princeton, NJ: Princeton University Press, 1984. A good survey of the new uniformitarianism.

Komar, P. D. "Comparisons of the Hydraulics of Water Flows in Martian Outflow Channels with Flows of Similar Scale on Earth." *Icarus* 37 (1979): 156–181. Bretz's vision is vindicated on Mars!

Malde, H. E. "The Catastrophic Late Pleistocene Flood in the Snake River Plain." *U.S. Geological Society Professional Paper* 506 (1968).

———. "Evidence in the Snake River Plain of a Catastrophic Flood from Pleistocene Lake Bonneville." *U.S. Geological Society Professional Paper* 440B (1960): B295–B297.

McKnight, E. T. "The Spokane Flood: A Discussion." *Journal of Geology* 35 (1927): 453–460. This work attacks Bretz's hypothesis of catastrophic flooding.

Meinzer, O. E. "Discussion: Channeled Scabland and the Spokane Flood." *Washington Academy of Science Journal* 17 (1927): 207–208. This work also attacks Bretz's hypothesis of catastrophic flooding.

Moore, John A. *Science as a Way of Knowing.* Cambridge, MA: Harvard University Press, 1993.

O'Conner, J. E. "Hydrology, Hydraulics, and Geomorphology of the Bonneville Flood." *Geological Society of America Special Paper* 274 (1993).

Pardee, J. T. "Unusual Currents in Glacial Lake Missoula, Montana." *Journal of Geology* 18 (1942): 376–386. This belated article gave Bretz a source for his floodwaters.

Shea, James. "Twelve Fallacies of Uniformitarianism." *Geology* 10 (1982): 455–460. An excellent and accessible article.

Webster, G. D., M. J. P. Kuhns, and G. L. Waggoner. "Late Cenozoic Gravels in Hells Canyon and the Lewiston Basin, Washington and Idaho." *Idaho Bureau of Mines and Geology Bulletin* 26 (1982): 669–683.

Chapter 7

Hallam, A. *Great Geological Controversies.* Oxford: Oxford University Press, 1983. This is an excellent history of many of the debates discussed in my book.

Schwartzbach, Martin. *Alfred Wegener: The Father of Continental Drift.* Madison, WI: Science Tech, 1986.

Wood, Robert M. *The Dark Side of the Earth.* London: Allen & Unwin, 1985. This is a witty, if one-sided, history of geology. It harshly judges some geologists earlier in the century while being extraordinarily kind to geophysicists and members of Princeton University's geology department.

Chapter 8

Kuppers, Bernd-Olaf. *Information and the Origin of Life.* Cambridge, MA: MIT Press, 1990. This book uses information theory to explore life's origins.

Lipps, Jere H. *Fossil Prokaryotes and Protists.* Oxford: Blackwell Scientific, 1993.

Mayr, Ernst. *Toward a New Philosophy of Biology*. Cambridge, MA: Harvard University Press, 1988. The grand dean of neo-Darwinism explains and defends his recent thinking. The book is not easy for the uninitiated.

Nelkin, D., and M. S. Lindee. *The DNA Mystique: The Gene as Cultural Icon.* New York: Freeman, 1995. This recent and accessible offering summarizes our culture's facination with DNA. Soon, the authors argue, almost every significant aspect of human life will be attributed directly to DNA. The absence of scientific data to support this trend has not, unfortunately, deterred journalists and other shapers of public opinion.

Ponnamperuma, Cyril, and Julian Chela-Flores, eds. *Chemical Evolution and the Origin of Life.* In *Proceedings of the Trieste Conference.* Ed. Cyril Ponnamperuma and Julian Chela-Flores. Hampton, VA: Deepak Publishing, 1993.

Weintraub, Pamala, ed. *The Omni Interviews.* London: Ticknor & Fields, 1984.

———. *The Search for Life's Origins.* Washington, DC: National Research Council, 1990.

Zubay, Geoffrey. *Origin of Life on Earth and in the Cosmos.* Dubuque, IA: Brown, 1996. This recent book is an excellent place to begin.

Chapter 9

Dott, Robert H., and Donald Prothero. *Evolution of the Earth.* 5th ed. New York: McGraw-Hill, 1995. This is a fine introduction to the physical and biological history of the earth.

Tattersall, Ian. *The Fossil Trail: How We Know What We Think We Know About Human Evolution.* New York: Oxford University Press, 1995.

Chapter 10

Barbour, I. G. *Issues in Science and Religion.* New York: Harper & Row, 1971. A good survey of many of the most contested issues.

Davies, Paul. *The Cosmic Blueprint.* London: Heinemann, 1987. A readable survey of the fundamental issues.

———. *The Mind of God: The Scientific Basis for a Rational World.* New York: Simon & Schuster, 1992. Both volumes are wonderfully written and entertaining.

Eldredge, Niles. *Reinventing Darwin.* New York: Wiley, 1995. This account of the issues from the view of an eminent paleontologist is recommended to students at all levels.

Gilson, E. *From Aristotle to Darwin and Back Again.* Trans. John Lyon. South Bend, IN: University of Notre Dame Press, 1984. This book provides a philosopher's perspective on these matters.

Mayr, Ernst. *One Long Argument.* Cambridge, MA: Harvard University Press, 1991. This book shows the matter from the neo-Darwinist perspective.

———. *Toward a New Philosophy of Biology.* Cambridge, MA: Harvard University Press, 1988. See especially p. 415.

Moorehead, Alan. *Darwin and the* Beagle. New York: Harper & Row, 1969.

Ruse, Michael. *Darwinism Defended.* Reading, MA: Addison-Wesley, 1982. This is a readable defense of neo-Darwinism.

Tassy, Pascal. *The Message of Fossils.* Trans. Nicholas Hartmann. New York: McGraw-Hill, 1993.

Teilhard de Chardin, Pierre. *Man's Place in Nature.* Trans. René Hague. New York: Random House, 1956.

———. *The Phenomenon of Man.* Trans. Bernard Wall. New York: Harper & Row, 1959. For the English reader, the books are difficult owing to the translation. Secondary reading is recommended.

Whittington, Harry B. *The Burgess Shale.* New Haven, CT: Yale University Press, 1985. As the Ediacara faunas do, the animals of the Burgess Shale indicate that the basic patterns of complex life on our planet could have been constructed much differently. Contingency rules all.

Chapter 11

Dennett, Daniel C. *Darwin's Dangerous Idea: Evolution and the Meaning of Life.* New York: Simon & Schuster, 1995. This is a readable summary of many issues from the viewpoint of neo-Darwinism.

Ehrlich, Paul R. *"Extinction: What Is Happening Now and What Needs to Be Done."* In *Dynamics of Extinctions*. Ed. David K. Elliot, pp. 157–164. New York: Wiley, 1986.

Eldredge, Niles. *The Miner's Canary: Unraveling the Mysteries of Extinction.* Englewood Cliffs, NJ: Prentice-Hall, 1991.

Gould, Stephen Jay. *Time's Arrow, Time's Cycle*. Cambridge, MA: Harvard University Press, 1987. This small volume is of interest to students of world civilization and biblical history as well as scientific imagery.

Hallam, Anthony. "End-Cretaceous Mass Extinction Event: Argument for Terrestrial Causation." *Science* 238 (1987): 1237–1241. Hallam is an anti-meteorite man, arguing that earth processes can account for what we see at the K–T boundary.

———. *Great Geological Controversies.* Oxford: Oxford University Press, 1989.

Lipps, Jere H. *Fossil Prokaryotes and Protists.* Oxford: Blackwell Scientific, 1993.

Mayr, Ernst. *Toward a New Philosophy of Biology.* Cambridge, MA: Harvard University Press, 1988.

Moore, John A. *Science as a Way of Knowing*. Cambridge, MA: Harvard University Press, 1993.

Raup, David M. *Extinction: Bad Genes or Bad Luck?* New York: Norton, 1991. A most engaging and accesible book, recommended to students at all levels.

Raup, David M., and J. J. Sepkoski. "Mass Extinctions in the Marine Fossil Record." *Science* 215 (1982): 1501–1503.

Simpson, G. S. "Uniformitarianism." In *Essays in Evolution and Genetics in Honor of Theodosius Dobzhansky.* Ed. M. K. Hecht and W. C. Steere, pp. 43–96. Englewood Cliffs, NJ: Prentice-Hall, 1970.

Stanley, Steven M., ed. *Extinction.* New York: Freeman, 1987. This is a good introduction to and overview of the subject.

Chapter 12

Field, A. N. *Why Colleges Breed Communists* (1941). Reprinted in 1984 as *The Evolution Hoax Exposed.* Rockford, IL: Christian Bookclub of America, Tan Books and Publishers. This is a weak link in the creationist chain of publications.

Ginger, Ray. *Six Days or Forever?* New York: Signet Books, 1958. This is an excellent and enjoyable account of the Scopes trial, better even than the movie *Inherit the Wind,* which also is fun.

Gish, Duane T. *Evolution: The Challenge of the Fossil Record.* El Cajon, CA: Creation-Life Publishers, 1985. This work is another example of creationism at its weakest. Duane Gish is a member of the Creation Science Institute in California.

Gould, Stephen Jay. "Justice Scalia's Misunderstanding." In his *Bully for Brontosaurus.* New York: Norton, 1991. Here Gould responds harshly to the dissenting opinion in the Arkansas "balanced treatment" law. He argues that evolution says nothing about life's origins, a position that may be true in the strictest sense but ignores what many evolutionary biologists in fact assume, teach, and write.

Johnson, Phillip E. *Darwin on Trial.* Madison, WI: Intervarsity Press, 1991. This is the best antievolutionist volume in print; it is on a plane wholly different from that of the creationist books and should be read by students at all levels with interest in this area.

Moreland, J. P., ed. *The Creation Hypothesis.* Madison, WI: Intervarsity Press, 1994. This is a serious attempt to launch a "theistic science," that is, a science that would treat the argument from design with the respect that some Christians think it deserves. Note that Intervarsity is an evangelical, conservative Christian publishing house.

Weiner, Jonathan. *The Beak of the Finch.* New York: Vintage Books, 1994. In my view, the basic claims of this book are radically overstated, but it won a Pulitzer Prize.

Chapter 13

Chisholm, R. M., and R. J. Swartz. *Empirical Knowledge: Readings from Contemporary Sources.* Englewood Cliffs, NJ: Prentice-Hall, 1973. Readable but challenging.

Gifford, N. L. *When in Rome: An Introduction to Relativism and Knowledge.* Albany: State University of New York Press, 1983.

Jahn, Robert G., and Brenda J. Dunne. *Margins of Reality: The Role of Consciousness in the Physical World.* New York: Harcourt Brace Jovanovich, 1987.

Nagel, Thomas. *The View from Nowhere.* Oxford: Oxford University Press, 1986. Nagel writes so well that his abstract topics flow smoothly through the mind.

Rescher, N. *Empirical Inquiry.* Totowa, NJ: Rowan & Littlefield, 1982.

Trigg, Roger. *Rationality and Science.* Oxford: Blackwell, 1993.

Wolterstorff, N. *Reason Within the Bounds of Religion.* Grand Rapids, MI: Eerdmans, 1976. This is a small and popular-level work. Recommended to beginners of all ages.

Index

Note:Page numbers in *italics* indicate illustrations.